CAPILLARY
MECHANICS

CAPILLARY MECHANICS

Shiqiao Gao
Beijing Institute of Technology, China

Lei Jin
Beiijing Institute of Technology, China

Deyi Fu
China Electric Power Research Institute, China

World Scientific

NEW JERSEY · LONDON · SINGAPORE · BEIJING · SHANGHAI · TAIPEI · CHENNAI

Published by

World Scientific Publishing Co. Pte. Ltd.

5 Toh Tuck Link, Singapore 596224

USA office: 27 Warren Street, Suite 401-402, Hackensack, NJ 07601

UK office: 57 Shelton Street, Covent Garden, London WC2H 9HE

Library of Congress Control Number: 2025009783

British Library Cataloguing-in-Publication Data
A catalogue record for this book is available from the British Library.

CAPILLARY MECHANICS

ISBN 978-981-98-1021-5 (hardcover)
ISBN 978-981-98-1022-2 (ebook for institutions)
ISBN 978-981-98-1023-9 (ebook for individuals)

For any available supplementary material, please visit
https://www.worldscientific.com/worldscibooks/10.1142/14222#t=suppl

Desk Editors: Nambirajan Karuppiah/Julio Hong/Amanda Yun

Typeset by Stallion Press
Email: enquiries@stallionpress.com

About the Authors

Gao Shiqiao, PhD, is a professor and doctoral tutor at Beijing Institute of Technology, China. He was awarded the Government Special Allowance of the State Council in 1997. He has been a research fellow of the Alexander von Humboldt Foundation since 1990. His current research interests include nonlinear structural dynamics, MEMS/NEMS technology and mechanics, and impact dynamics. He has published more than 150 papers in many important international and domestic academic journals and authored eight monographs.

Jin Lei is currently an associate professor at the School of Mechatronic Engineering, Beijing Institute of Technology, China. Her current research interests include electromechanical and control technology, microelectromechanical technology, and micro energy technology. She has published over 30 papers in academic journals and participated in the writing and publication of six monographs.

Fu Deyi, PhD, is a professorate senior engineer and senior technical expert at China Electric Power Research Institute (CEPRI). His research interests include wind turbine measurement and certification, structural dynamics analysis and control, fault diagnosis and monitoring, and fatigue assessment. He is the leader of the remote sensing expert group of the Measuring Network of Wind Energy Institutes (MEASNET), a member of the power performance and mechanical load expert group of MEASNET, a member of IEC TC88 MT13, a peer assessor of IEC RE wind energy, and also a member of the SG551 expert group. He also serves as a master supervisor at Beijing Institute of Technology (BIT) and North China Electric Power University (NCEPU). He is a nationally registered metrologist. He has published over 30 papers, obtained 16 authorized patents, and 11 software copyrights.

Contents

Introduction

In recent years, micro/nano science has been continuously developing, and micro/nano technology has been continuously advancing. Numerous high-tech concepts and tangible achievements have emerged in materials, structures, and systems. Some of them have been applied in industrial and information-based commodity production, while others have great application prospects and potential.

As the micro/nanoscale ranges from micrometers (10^{-6} meters) to nanometers (10^{-9} meters), it is not only different from the traditional macroscopic scale visible to the naked eye above millimeters but also different from the molecular atomic scale invisible to the naked eye at the traditional physical and chemical levels. Therefore, for a long time, its role has been overlooked at the micro level, and its existence has been overlooked at the macro level. It was referred to as the "neglected size world" by Wolfgang Ostwald. However, such a size is the transition size from macro to micro, the connection size between macro and micro, and also the interaction size of interfaces and surfaces. At this size scale, it involves many production and daily life issues. The connection of interfaces, surface adsorption, interface adhesion, mechanical lubrication, friction, wear, surface pollution and cleaning, surface corrosion, etc., are all related to the effects of this scale. Especially when the size of machinery and structures itself is reduced to this level, the interaction between different phases or states of matter becomes more significant. Many of these issues involve capillary action. Therefore, it is worth studying and gaining a deeper understanding.

Capillary action results from the influence of intermolecular forces at the microscopic level and manifests as a macroscopic visible phenomenon, thus spanning a large scale. Traditional capillary theory is mostly based on macroscopic phenomenological research, such as thermodynamic theory, energy theory, and fluid mechanics theory. Even if it involves microscopic aspects, it only explores the influence of highly microscopic molecular forces on the mechanism of capillary phenomena from a qualitative perspective, rarely involving the scale of micro and nano. However, due to the fact that capillary action itself involves three "phases" or "states" of gas, liquid, and solid and mainly involves surface and interface interactions, its effect is strong at the micro and nano scales. At present, there is still a lack of specialized English books like this. From this perspective, there is both a need for a re-understanding of capillary phenomena and capillary effects, as well as a need for in-depth research. This book is written from this perspective.

Capillary mechanics can be traced back to ancient times. As early as the 13th century, people proposed the concept of capillaries and attempted to understand the laws of capillary action from the perspective of blood circulation.

Capillary mechanics is also very modern. Until the 21st century, people have been constantly exploring capillary effects in different fields, such as capillary effects in micro and nano fields, and capillary effects in surface science.

Capillary mechanics is very traditional. This science was born almost contemporaneously with traditional Newtonian mechanics and developed alongside it. Many laws are also based on Newtonian mechanics.

Capillary mechanics is also very fashionable. Many new interdisciplinary fields in contemporary times are either derived from or closely related to capillary mechanics, such as nanoscience, Micro-Electro-Mechanical systems (MEMS) science, materials science, and interface chemistry.

The most basic principles of capillary mechanics are simple and straightforward. The most fundamental concept of capillary mechanics is surface tension, or surface free energy. The explanation of capillary phenomena is often based on surface tension or surface free energy.

The laws of capillary mechanics are extremely complex and profound. Due to the fact that molecular-level forces determine surface tension or surface free energy properties and the complexity of the interaction patterns between a pair of molecules, it is necessary to explore the interactions between a group of molecules, whether from the perspective of non-vector energy or vector force, which are extremely complex. Not only is the physical process extremely complex, but the mathematical form is also extremely complex. Even from a phenomenological perspective, the laws governing the action of a general form of matter are complex.

The capillary phenomenon is very common and also very usural. Insert a thin glass tube into a liquid, such as water, and you will see the liquid rise inside the glass tube. By dipping a straw in a soap solution, you can blow out colorful bubbles, and children can conduct such experiments.

The capillary phenomenon, however, contains extremely rich scientific principles. From the burning of wick oil, the absorption of water by sponge, to the growth of crops and plants, soil and water conservation, artificial rainfall in astronomy and meteorology, all involve basic capillary action.

The capillary action is very microscopic. Its mechanism of action belongs to the action of intermolecular forces. And the molecules are only in the size range of 0.1 nm, and the molecular spacing is also only in the range of 0.1 nm. Therefore, the origin of their effects belongs to microscopic processes.

The capillary action is also very macroscopic. The capillary phenomenon caused by the microscopic effects of capillaries is actually macroscopic. Some may have a slow speed, but the cumulative effect is visible to the naked eye at a macroscopic level, while others exhibit capillary phenomena, and their macroscopic speed is quite fast. When a nurse pricks the finger of an individual to take a blood sample, it can clearly be seen that the blood is moving at a faster speed in a thin test tube with the naked eye.

The capillary action is very small. Liquids can only exhibit capillary rise in very thin tubes. And generally, its upward speed is also very slow.

The capillary action is also grand. The so-called high mountains and deep waters depend on the action of capillary force. The Earth nourishes all things and grows through the action of capillaries, while the vapor of mountain clouds and the transformation of clouds into rain also rely on the action of capillaries.

The capillary phenomenon is often utilized by people in their daily lives. The tissues that wipe sweat, the corduroy sweatshirts that absorb sweat, and the sponges that wipe the floor, all of them utilize capillary action.

The capillary phenomenon has also attracted more attention from ancient and modern scientists. Leonardo da Vinci of the Renaissance period focused on capillary phenomena, Sir Isaac Newton studied capillary phenomena, and Thomas Young and Pierre Simon Laplace further promoted the development of capillary mechanics. In addition, C.F. Gauss, Johann Bernoulli, and Simeon Denis Poisson have also made many contributions to the development of capillary mechanics. James Thomson, Marangoni, and others further developed capillary mechanics. Lord Kelvin and others were a groundbreaking driving force in the development of capillary mechanics. Moreover, even the first paper by the greatest modern physicist, Albert Einstein, explored physical problems by studying and utilizing capillary phenomena.

There are many reasons why capillary phenomena and their effects have attracted the attention of scientists from ancient and modern times, both at home and abroad. The above characteristics are also part of the reasons, but four of them are the most fundamental. One reason is that it is the link and bridge between physical macro and micro interactions. The capillary phenomenon is a unity of microscopic effects and macroscopic characterization. Due to the limited level of technology in the previous centuries, people were unable to directly observe the underlying phenomena, especially the molecular-level effects that could not be measured from a physical perspective. However, due to the macroscopic phenomenon of capillaries, it can be observed with the naked eye. Therefore, through macroscopic observation and measurement of capillary phenomena, one can indirectly understand the laws of the microscopic material world. The second reason is that it simultaneously involves different forms of matter, whether from the perspective of "phase" or "state". The gas, liquid, and solid in the material world are all involved simultaneously, making it both typical and representative, as well as reflective of the generality of the world's materials, especially the composition of objects. The third reason is that the core of capillary action is the effect of interfaces. Compared to the ontology, the interface has both step and real variability, as well as transitional and continuous properties. Therefore, its mechanism of action is relatively complex. There are both stable and unstable trends, with both physical and chemical effects. Therefore, the study of capillary phenomena directly gave birth to modern interface science and surface science, which are the foundations of science and technology, such as connection, lubrication, friction, cleaning, wear, adhesion, and adsorption. The fourth reason is that capillary phenomenon is widely present in nature and the actual production and life of people. And its underlying principles are complex and profound. Universality determines people's attention, while complexity and depth attract the attention of scientists.

From the above, it can be seen that capillary phenomena have always been familiar to people, and capillary action has also been the focus of scientists. However, unfortunately, capillary mechanics has not yet formed an independent discipline. In modern times, they are only a branch of thermodynamics and fluid mechanics and a part of surface science.

However, in the 21st century, micro/nano technology has emerged and attracted global attention. Major countries around the world regard micro/nano technology as the most likely scientific and engineering field to achieve breakthroughs in the future. The rise of micro/nano technology has unprecedentedly increased human interest, cognitive level, and desire to manipulate the world of micro/nano technology. The continuous development of micro/nano-electro-mechanical technology and the continuous miniaturization of micro/nano-mechanical structures have prompted people to constantly explore deep-seated problems, and capillary mechanics has thus been revitalized.

With improvements in manufacturing technology, the scale of research objects continues to shrink, and the ratio of surface area to volume of materials or gaps increases relatively. The importance of surface force (or interfacial force) relative to volume force is greatly enhanced, resulting in a significant increase in the importance of capillary action. The importance of capillary action at the micrometer scale is an undeniable fact, and the importance at the nanoscale is certainly self-evident.

At the micro/nanoscale, capillary action is a double-edged sword. On the one hand, it has great usability, for example, people can use capillary action to perform "capillary casting", make "molecular straws", and "nano test tubes" and also perform self-assembly of micro-devices. Capillary action has been successfully used to guide and control fluid flow in microfluidic actuators, and people naturally expect it to play a stronger role in nanoscale fluid manipulators. On the other hand, at the micro/nanoscale, capillary action can bring many unfavorable challenges that must be addressed. The capillary adhesion effect can lead to "adhesion" or even "failure" between the active parts of MEMS and even more severe "adhesion" or "failure" between the active parts of nano-mechanical systems (NEMS).

It is during the continuous exploration process that people have found that capillary mechanics has a particularly prominent impact on the performance of micro/nano-mechanical structures, so research on micro/nanoscale capillary interactions has received great attention. However, people's understanding of capillary mechanics is still limited, and there is a lack of sufficient understanding of many important issues. Many deep-seated problems have not been well solved. In order to better promote the study of capillary action, it is necessary for capillary mechanics to become a relatively independent science and to combine with the development of micro-nano technology and science, forming a modern research field that is full of new vitality.

Chapter 1

Capillary and Capillary Action

Capillarity is a prevalent physical phenomenon observed in nature. The ancient Chinese adage "When the clouds rise, the plinth turns wet. When the plinth becomes wet, rain will soon follow" illustrates the capillary phenomenon in the natural world. Most plinths are composed of porous materials. When the humidity in the air is elevated, it contains significant amounts of moisture. The process of capillary condensation can lead to the saturation of plinths, serving as an indicator that rainfall is imminent.

1.1 The Phenomenon and the Function of Capillary

Initially, we will examine two intriguing phenomena to enhance our understanding of capillarity.

One can place a drop of mercury on a clean glass surface. The drop will roll off without adhering to the glass. Similarly, if a clean glass pane is immersed in mercury and subsequently removed, the mercury will not cling to the surface of the glass. This phenomenon, characterized by the inability of a liquid to adhere to solid surfaces, is referred to as non-wetting. Mercury exemplifies a non-wetting liquid in relation to glass.

When a drop of water is placed on a clean pane of glass, it adheres to the surface and forms a thin layer. Similarly, when a clean pane of glass is immersed in water, the surface becomes coated with a layer of water. This phenomenon, in which a liquid adheres to the surface

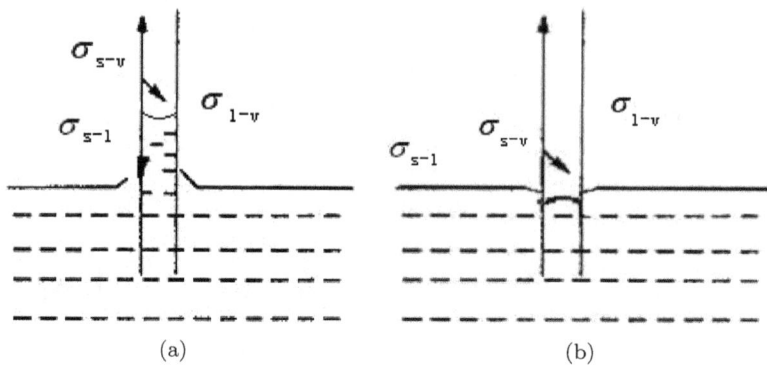

Figure 1.1. The capillary tubes dip into liquids. (a) Glass tube immersed into water and (b) glass tube immersed into mercury.

of a solid, is referred to as wetting. Water is classified as a wetting liquid in relation to glass.

The ability of a liquid to wet a particular type of solid may not extend to other types of solids. For instance, water is capable of wetting glass, whereas it does not wet paraffin. Mercury is unable to wet glass but is capable of wetting zinc.

When a superfine glass tube is immersed in water, the water level within the tube rises. The smaller the inner diameter of the tube, the higher the water ascends (see Figure 1.1(a), where "s" denotes solid, "l" denotes liquid, and "v" denotes vapor). Conversely, when these tubes are submerged in mercury, the mercury level within the tube decreases. Again, the smaller the inner diameter of the tube, the lower the mercury level drops (see Figure 1.1(b)). This phenomenon, characterized by the ascent of a wetting liquid and the descent of a non-wetting liquid within a superfine tube, is referred to as the "capillary phenomenon". Tubes that exhibit a pronounced capillary phenomenon are known as capillary tubes. Examples of materials that contain capillary tubes include paper, lamp wicks, gauze, soil, and the roots and stems of plants. The liquid surface within the tube can present either a concave or convex meniscus. When the liquid wets the solid material of the capillary tube, the liquid surface rises and forms a concave shape. Conversely, when the liquid does not wet the solid material, the liquid surface drops and takes on a convex shape.

The capillary phenomenon is prevalent in both industrial production and everyday life, with numerous illustrative examples.

Calcareous tufa, which is characterized by the presence of numerous capillary tubes, can be utilized as an artificial hill, akin to a bonsai, facilitating moisture acquisition for the plants situated on this artificial structure. This moisture ascent is fundamentally dependent on the principles of capillarity. In plants, the vascular structures within the roots and stems consist of extremely fine capillary tubes. The capacity of plants to absorb moisture and nutrients from the soil through their roots and stems is significantly attributed to the capillary phenomenon and the functionality of these capillary tubes. Additionally, the lubricating oil in machinery permeates the interstitial spaces of machine components, providing lubrication through the action of capillary tubes. Furthermore, various porous solid materials, such as bricks, paper, textiles, and chalk, exhibit the ability to absorb water due to the abundance of minute pores within these materials, which function as capillary tubes.

The capillary effect plays a significant role in water conservancy, facilitating the movement of groundwater from areas of higher moisture to those of lower moisture. This phenomenon serves as a vital source of water for plants. Additionally, capillary action enables facial tissues to absorb liquids when used to wipe surfaces. Each pore in a sponge functions similarly to a capillary tube, allowing the sponge to absorb a substantial volume of liquid. Furthermore, traditional corduroy sportswear effectively draws sweat away from athletes' skin by employing the principle of capillary action, akin to that of a lamp wick.

In certain circumstances, capillary action can be detrimental. For instance, the presence of excessively thin capillary tubes in the foundation of buildings can lead to the upward movement of soil moisture, resulting in increased indoor humidity. To mitigate this issue, asphalt felt is often applied to the base as a preventive measure against moisture intrusion caused by capillary action. Additionally, the phenomenon of water rising through capillary tubes can significantly adversely affect agricultural production. The soil contains numerous capillary tubes that facilitate the upward movement of moisture from below the surface. To conserve subsurface moisture, it is essential to aerate the soil and disrupt the capillary tubes at the surface, thereby reducing moisture evaporation.

Every phenomenon possesses dual aspects. A collaborative effort involving the University of Paris, the University of ESPCI, and

École Polytechnique seeks to transform this detrimental effect into a beneficial application. This research group aims to manipulate micro-plates to form three-dimensional shapes utilizing capillary action. The process unfolds as follows: thin silicone is sculpted into various configurations, such as flowers, triangles, or squares, after which a droplet of water is applied to the surface. Due to the water's tendency to minimize its contact area with the surrounding air, the droplet promptly begins to conform to the contours of the thin plate. Following the evaporation of the water at room temperature, the curved structure gradually solidifies, ultimately achieving its intended shape. Notably, this entire process occurs spontaneously, requiring no further intervention after the initial application of the droplet.

In summary, capillarity is not only prevalent in nature and daily life, but it also demonstrates both beneficial and detrimental effects. Understanding how to harness this phenomenon to maximize its advantageous properties while mitigating its adverse effects is crucial for the development of Micro-Electro-Mechanical Systems (MEMS).

1.2   The Capillary Action of M/NEMS

The aforementioned capillary phenomenon is observable at a macroscopic level. While certain microscopic processes associated with capillary action may not be directly observed, the resultant macro effects can still be quantified. It is evident that the thinner the capillary tube, the more pronounced the capillary action becomes. Similarly, an increase in the number of capillary tubes correlates with a heightened significance of capillary action. With advancements in Micro/Nano-Electro-Mechanical Systems (M/NEMS) technology, the dimensions of transformable microstructures or mobilizable components have reached the micron and even nanometer scales. In contrast to traditional macro functions generated by numerous capillary tubes, the individual capillary action is markedly evident in M/NEMS applications. For instance, during the fabrication of surface micro-mechanical structures, enhanced capillary interactions often cause micro-bridges and micro-trabecular structures to adhere to the substrate, thereby compromising their efficacy. Additionally, in experiments assessing material constants at the micro-scale, micro-bridges and micro-beams are frequently utilized as specimen structures for performance testing. The presence of capillary adhesion between the loading apparatus and the testing microscale specimen can significantly affect the accuracy of the obtained data. In micro-surface processing, when the sacrificial layer is completely etched, deionized water is required to remove etchants and residual materials. However, this process can easily lead to the formation of a liquid bridge interface between two parallel surfaces upon removal from the deionized water. Such liquid bridges generate substantial capillary forces. Notably, even in the absence of a liquid bridge during manufacturing, humidity can facilitate the formation of liquid bridges within micro-structures. The emergence of liquid bridges is a direct consequence of capillary action, which in turn generates relevant liquid bridge forces. For MEMS, inadequate packaging can lead to the formation of capillary condensate at relative humidity levels of 65%. Under the influence of capillary forces, if the restoring force of the structure is insufficient, microscale bridges may adhere to the substrate. Therefore, it is imperative to comprehend the emergence and function of capillary forces and to effectively manage and utilize these forces in practical applications.

The advancement of machining precision has progressed from micron dimensions to the nanometer scale. Currently, the precision of ultra-precision diffraction gratings has achieved a remarkable accuracy of 1 nm, facilitated by diamond turning techniques. Furthermore, laboratories have successfully fabricated structures such as wires, columns, and troughs measuring less than 10 nm. With the rapid evolution of M/NEMS process technology, M/NEMS devices have gained widespread application. The capillary phenomenon is present across all scales; however, its significance is particularly pronounced in M/NEMS. The presence of capillarity plays a crucial role in both the production and operational processes of these devices.

Most designs of M/NEMS typically incorporate several fundamental components, including micro- and nanoscale springs, cantilever beams, films, hinges, and gears. During the manufacturing processes of these standard structures, the final step at the wafer level involves the release of all functional components. This process commonly entails the etching of a sacrificial layer. Once the sacrificial layer is removed, the sections composed of permanent materials are allowed to move freely in accordance with the intended design. A critical aspect of the release process is the thorough cleaning of the structures to eliminate etchants, dissolved silica (or other sacrificial materials), microscale solid residues, and any potential remaining by-products of the sacrificial layer. However, a significant challenge arises during this process: the flexible sections may experience capillary adhesion due to capillary forces. When the force of capillary adhesion exceeds the mechanical restoring force of the microstructure, adherence occurs. As liquid comes into contact with adjacent surfaces and evaporates slowly, the capillary force generated by the liquid detergent can cause several sections to adhere to one another. This phenomenon can be attributed to surface tension. In the context of microfabrication technology, once the sacrificial layer is removed and the structure is released, the structural layer is prone to adhere to the substrate or neighboring structures. This occurrence is referred to as release adhesion.

The presence of moisture is generally detrimental to electronic devices, particularly to M/NEMS. During operation, if M/NEMS devices are inadequately encapsulated, capillary condensation can lead to the formation of liquid droplets due to the hydrophilic nature of the materials used in these devices under conditions of high humidity. Such droplets can create liquid bridges within structures, such as micro-resonators, micro-accelerometers, and micro-gyroscopes, which in turn generate resistance or adhesive forces that impede the relative motion of the structures, ultimately resulting in device failure. This phenomenon is referred to as working adhesion. Recent research conducted by Sandia National Laboratories has indicated that, once the operational conditions meet the strength requirements of M/NEMS materials, the primary cause of device failure is the capillary adhesion of micro/nano-mechanical structures. For instance, if a liquid bridge due to capillary adhesion forms between the electrodes of a micro-switch, the contact angle of the meniscus and the surface roughness of the substrate will significantly affect the pull-in voltage. In experiments involving micro- and nanoscale structures, micro/nano bridges and micro/nano girders are commonly utilized as test specimens for assessing material constants and functions at the microscale. The presence of capillary adhesion between the microscale testing end and the loading end can directly impact the accuracy of the test data and may even lead to complete failure of the test specimen.

The significance of capillarity in M/NEMS is influenced by several factors, including the diminutive size, relatively large surface area, and the close proximity of adjacent surfaces in micro- and nanoscale devices. The volume forces associated with mass and quality, such as gravitational and magnetic forces, diminish rapidly with the cube of the dimension, whereas surface forces related to surface area, including frictional, adhesive, and van der Waals forces, decrease with the square of the dimension. Consequently, in micro-devices, volume forces become negligible, while surface forces emerge as predominant factors. As a result, with the miniaturization of components, the impact of volume forces, such as gravity, can often be disregarded; however, the surface forces arising from contact or friction between surfaces assume a critical role. The adhesion among structural elements significantly affects the functionality and reliability of micro and nano machines. Given the importance of capillary adhesion in the architecture of micro- and nanoscale devices, research into the mechanical mechanisms of capillarity at the micro and nano scale has proliferated in recent years.

Capillary action frequently manifests as a liquid bridge in M/NEMS that involve mechanical and electrical components. This phenomenon occurs when two solid surfaces are brought sufficiently close together, allowing a small volume of liquid to be absorbed between them. The liquid bridge phenomenon is prevalent in various contexts and induces oligodynamic interactions between the solid surfaces and the liquid. To separate the two solid surfaces, a force must be exerted that exceeds the capillary force exerted by the liquid bridge. This force is commonly referred to as the liquid bridge force or capillary force.

In micro- and nanoscale systems, the phenomenon of liquid bridging is observed among the structures of micro and nano devices. For instance, a liquid bridge can form between a cantilever beam and its substrate under conditions of high relative humidity, as seen in miniature accelerators (or RF MEMS). Similarly, small droplets of water can create a liquid bridge between the comb tines of a micro-gyroscope and its base. This liquid bridge induces a capillary adhesion effect between the solid surfaces and the liquid. At the micro and nano scales, surface forces become significantly more influential than volumetric forces. The capillary adhesion effect can adversely affect the normal operation of RF MEMS,

micro-transducers, micro-gyroscopes, and other devices characterized by micro- and nanostructures. The liquid present in the liquid
bridge may either be pre-existing or may form gradually from nearly
saturated condensable vapor. Consequently, in microscopic systems,
particularly those with nanostructures, the presence of original liquid
increases the likelihood of liquid bridge formation under conditions
of elevated moisture. To mitigate the formation of liquid bridges in
microscopic or nanoscale systems, it is essential to incorporate ade-
quate vacuum desiccants during the packaging process.

The liquid bridge is a prevalent phenomenon in micro- and
nanoscale systems, particularly in measurement techniques and
equipment designed for micron- and nanosized applications, where
the liquid bridge exhibits significant effects. Currently, measurement
techniques for geometric characteristics and surface topography with
micro- and nanoscale accuracy have been developed. For instance,
advancements in Scanning Tunneling Microscopy (STM), Scanning
Probe Microscopy (SPM), and Scanning Force Microscopy (SFM)
have enabled cutting-edge operations, fabrication, and reshaping at
the atomic level in recent years.

The operational principles of measuring devices such as STM,
SPM, and SFM indicate that these instruments are susceptible to
the effects of capillary action. The SPM is fundamentally based on
the tunneling effect as described by quantum mechanics. When the
atomic-scale tip of the probe scans samples at a height of less than
1 nm, the electron clouds of the tip and the sample overlap. By applying a voltage ranging from 2 mV to 2 V, the tunneling effect facilitates
the escape of electrons from the sample, resulting in the formation of
a tunneling current. The intensity of this current exhibits a functional
relationship with the distance between the probe tip and the sample surface. Due to the uneven nature of the atomic structure at the
sample's surface, as the probe tip traverses the material at a specified
height, the distance between the tip and the surface varies continuously, leading to corresponding fluctuations in the tunneling current.
The graphical representation of these current variations can reveal
atomic-level surface patterns. In addition to the STM, a variety of
SPM techniques have been developed that exploit different interactions between the probe tip and the samples to investigate surfaces
or interfaces at the nanoscale. These advanced probe microscopes
include Atomic Force Microscopy (AFM), Laser Force Microscopy
(LFM), and Magnetic Force Microscopy (MFM), all of which are

derived from the principles of STM. The AFM, for instance, operates by having a microprobe glide across the sample surface, causing a highly sensitive microcantilever beam to oscillate in response to the surface topography. The AFM detects the displacement of the micro-cantilever beam using optical methods or tunneling current, thereby measuring the repulsive forces between the tip's atomic point and the surface atoms. Given the operational principles of probe microscopy, there exists a significant likelihood of liquid bridges forming between the microscope tip and the surface of the object being measured. If small droplets of liquid are present between the microscope tip and the sample, a liquid bridge may be established. Additionally, films of water or other liquids can also create liquid bridges. The presence of these liquid bridges can adversely affect the measurement accuracy of these microscopy techniques.

Currently, capillary action presents a significant challenge in the research and development of M/NEMS devices that has yet to be addressed. Effectively controlling capillary action during the manufacturing of micro and nano devices has emerged as an issue that warrants considerable attention.

1.3　The Development of Capillary Mechanics

The term "capillary" comes from the well-known concept "blood capillaries". Ibn al-Nafis (1213–1288), a prominent Arab scientist and physiologist, was the first to propose the idea that nutrients are obtained from the pulmonary artery through capillary circulation in the 13th century. However, at that time, his understanding of the mechanisms governing blood flow within capillaries was limited. It was not until the years 1845–1846 that German physicist Gotthilf Heinrich Ludwig Hagen (1797–1884) and French physiologist Jean Louis Marie Poiseuille (1797–1869) conducted research that established the principles governing blood flow in capillaries, building upon the general understanding of capillary action. Although the concept of capillarity originates from blood capillaries, the phenomenon of capillary action encompasses a variety of occurrences in nature. In fact, capillary phenomena and their effects are prevalent in both the natural world and everyday life.

Although the concept of capillarity was introduced in the early 13th century, it was not until the 18th century that systematic studies of capillarity began to emerge, culminating in a more comprehensive understanding by the late 19th century. During this period, several prominent scientists made significant contributions to the advancement of capillary theory. Their innovative work propelled the development of both physics and capillary mechanics. Consequently, several key theories and concepts in capillary mechanics are named in honor of these scientists, serving as a tribute to their pioneering contributions. Notable examples include the Young–Laplace equation, Gibbs phase rule, Kelvin equation, Helmholtz free energy, and Marangoni effect.

According to the German physicist Johann Christian Poggendorff (1796–1877), the capillary phenomenon was first discovered by Leonardo da Vinci (1452–1519), who is recognized as one of the prominent figures of the Italian Renaissance. Leonardo da Vinci exhibited a broad interest in research and investigated numerous significant phenomena in nature. However, the earliest documented discovery of the capillary phenomenon is attributed to the English scientist Francis Hawksbee (1666–1713). In 1709, Hawksbee was the first to observe the capillary phenomenon in tubular structures and on glass plates, noting the attraction between the glass surface and the liquid. He found that the capillary phenomenon occurs in both thick and thin glass tubes and concluded that its influence is present only when the liquid comes into contact with the internal surface of the glass tube.

In 1718, the English scientist James Jurin (1684–1750) demonstrated that the height to which a liquid rises is solely a function of the liquid's surface area. Additionally, he established that the height of liquid in capillary tubes is inversely proportional to the radius of those tubes. Jurin's significant contributions to the understanding of capillary action garnered him widespread recognition in the scientific community.

The English mathematician, physicist, and astronomer Sir Isaac Newton posited that the phenomenon of capillarity is influenced by the viscosity of liquids during his investigations into this property. Nevertheless, he did not provide a comprehensive explanation of the characteristics of capillary action in liquids.

In 1751, the German scientist Johann Segner (1704–1777) first introduced the significant concept of surface tension in liquids, attributing this phenomenon to gravitational interactions. He endeavored to demonstrate that surface tension influences the shape of liquid droplets. While Segner acknowledged the effect of radial curvature on liquid droplets, he overlooked the impact of horizontal curvature. His contributions to the understanding of surface tension enhanced the comprehension of capillary action.

In 1756, the German physician Johann Gottlob Leidenfrost (1715–1794) noted that soap bubbles exhibited a tendency to shrink. He observed that when the inlet side of the tubule connecting to the soap bubble was opened, the bubble would continuously contract in an effort to expel the air. However, Leidenfrost did not ascribe this phenomenon of shrinking to the effects of the liquid surface.

In 1787, the French mathematician and chemist Gaspard Monge (1746–1818) applied the principle of surface tension to elucidate the evident attractive and repulsive forces that exist between liquids and the objects that float upon their surfaces.

In 1802, the Scottish mathematician and physicist John Leslie (1766–1832) provided the first accurate explanation of the rise of liquid in capillary tubes due to capillary action. He did not assume that the direction of the attractive force between the tube and the liquid was upward, which would facilitate the movement of the liquid. Instead, Leslie emphasized that the primary effect of the attractive force was to increase the pressure of the fluid layer in contact with the tube, which was significantly greater than the internal pressure of the liquid. This phenomenon can be intuitively understood as causing the fluid contact layer to spread across the surface of the solid. Consequently, when a droplet of liquid is placed on a clean, flat glass surface, it tends to spread out. This explains why a glass tube becomes wet when immersed in water. The rise of the liquid continues until the weight of the water column is balanced by the surface tension forces, at which point the liquid ceases to rise. Leslie's theory was subsequently validated through mathematical methods by the Scottish mathematician James Ivory (1765–1842).

Thomas Young (1773–1829), an eminent English physicist, established the theory of capillarity, which is grounded in the concept of surface tension, in 1804. Young recognized that the contact angle between a liquid's surface and a solid substrate remains constant, and he elucidated the phenomenon of capillary action based on two foundational principles. In 1805, he was the first to propose the equations governing solid surface tension, liquid surface tension, and solid–liquid surface tension, collectively known as Young's equation. This equation serves as a fundamental formula in the analysis of the wetting phenomenon within the framework of surface phenomena, and it is often referred to as the "wetting equation". Thomas Young was a polymath in Britain and a key figure in the development of the wave theory of light. His research spanned multiple disciplines, including capillary action, light wave theory, sound wave theory, hydrokinetics, shipbuilding engineering, tidal theory, the measurement of gravity using pendulums, and the theory of rainbows.

Thomas Young and the French mathematician Pierre-Simon Laplace (1749–1816) independently derived the relationship between the additional pressure exerted beneath the curved surface of a liquid and the radius of curvature, a principle known as the Young–Laplace Equation, or simply the Laplace equation. Pierre Simon Laplace was a distinguished mathematician, astronomer, and member of the French Academy of Sciences. He was elected as a member of the Collège de France in 1816 and subsequently served as its president in 1817. Laplace was a principal figure in the development of celestial mechanics, a co-founder of cosmogony, and a pioneer in the field of probability theory. Although the research findings of Laplace in capillary mechanics were remarkably similar to those of Thomas Young in several respects, their methodologies were fundamentally different. Laplace arrived at significant conclusions primarily through mathematical calculations.

The research on capillarity progressed to a more significant phase during the early 19th century. Carl Friedrich Gauss (1777–1855), a prominent German scientist, introduced the method of energy analysis in 1830. Gauss, renowned as a mathematician, physicist, astronomer, and expert in geodesy, has been posthumously enshrined in academic discourse as "the Prince of Mathematics" — an honorific epithet denoting his unparalleled eminence in mathematical sciences. He derived various equations related to capillary action and established boundary conditions based on the principles of energy conservation and virtual work, the latter of which was articulated by Johann Bernoulli (1667–1748), a Swiss mathematician. In his formulation, Gauss employed a concept now recognized as potential energy. He concluded that potential energy comprises three components: gravitational potential energy, interaction energy among fluid particles, and the potential energy associated with the interaction between the fluid and the solid contact surface. Furthermore, Gauss identified deficiencies in the work of Pierre-Simon Laplace.

In 1831, Simeon Denis Poisson (1781–1840), a prominent French mathematician, published a monograph addressing the phenomenon of capillary action. He posited that the density of a liquid near its surface undergoes rapid changes. This assertion was subsequently validated through a series of experimental investigations. Poisson conducted research on fluid equilibrium based on the hypothesis of uniform density, ultimately concluding that the capillarity observed under this assumption was inconsistent with theoretical predictions. Consequently, Poisson demonstrated that Laplace's hypothesis regarding uniform density was flawed. However, it is important to note that Laplace's hypothesis, which suggested that liquid density was uniform and that attractive forces existed among liquid molecules, proved effective within a limited scope. Furthermore, Gauss later corroborated the conclusions derived from this hypothesis through alternative methodologies. As a result, Poisson's conclusions were ultimately deemed incorrect. Nevertheless, Poisson's work laid the groundwork for further investigation into the varying densities of fluids at the surface compared to those within the bulk. In essence, Poisson's study was fundamentally equivalent to Laplace's conclusions. Both fluid surface equations necessitate the introduction of a constant, denoted as H, which must be determined experimentally. The primary distinction lies in the fact that the constant H is contingent upon either the principles of molecular forces or the principles of surface density of the fluid. Currently, it remains challenging to differentiate between these two principles.

James Thomson (1822–1892), the elder brother of Lord Kelvin and an English physicist, first observed the phenomenon known as the tears of wine in 1855. The Italian physicist Carlo Marangoni (1840–1925) completed a doctoral dissertation on this phenomenon in 1865, leading to its designation as the Marangoni effect. Much of the subsequent research on this topic was conducted by the American physicist Josiah Willard Gibbs (1839–1903), which has resulted in the phenomenon also being referred to as the Gibbs–Marangoni effect. The tears of wine in wineglass are related to the presence of a surface tension gradient, which induces liquid flow toward regions of higher surface tension.

Inspired by Gauss's energy theory, numerous subsequent studies have been conducted that analyze phenomena not only from the perspective of force balance but also from the standpoint of systematic energy conservation. To date, there exist two definitions of surface tension: one derived from energy considerations and the other from force considerations, both of which are fundamentally equivalent. Each of these approaches has its respective advantages and disadvantages. The force-based method is intuitive and straightforward; however, it is limited by the vector nature of force, which introduces directional considerations that complicate the analysis of complex systems. Conversely, the energy-based method circumvents the issues associated with vector directionality and allows for arithmetic summation. Nonetheless, this method is less effective for analyzing interactions and may not be as intuitively comprehensible. Thermodynamics is fundamentally a phenomenological discipline grounded in energetics. Given that capillary phenomena are typically regarded as a subset of thermodynamics, the analytical framework of energetics has significantly influenced advancements in capillary mechanics. Notably, Josiah Willard Gibbs and Hermann von Helmholtz are two of the most prominent and influential figures in the field of thermodynamic energetics.

Gibbs established a law in 1876 that describes the relationship among the number of phases, the number of components, and the degrees of freedom in an equilibrium system. This principle, known as the phase rule, articulates a universal law governing phase equilibrium. Gibbs free energy, also referred to as the Gibbs function, is a crucial parameter in thermodynamics, commonly denoted by the symbol G. It is defined as the maximum work obtainable from a thermodynamic system during isothermal and isobaric processes, excluding work associated with volumetric changes. In other words, during isothermal and isobaric processes, the work performed by the system on its surroundings, apart from that due to volumetric changes, can be equal to or less than the decrease in Gibbs free energy. Gibbs made significant contributions to both thermodynamics and statistical physics, and his work has led to the establishment of a rigorous and comprehensive theoretical framework in thermodynamics.

Hermann von Helmholtz (1821–1894), a prominent German physicist and physiologist, published an influential article titled *Thermodynamics in Chemical Processes* in 1882. In this work, he distinguished between binding energy and free energy in chemical reactions, noting that binding energy can only be converted into heat, whereas free energy can be transformed into various other forms of energy. Helmholtz derived what came to be known as the Gibbs–Helmholtz equation from the Clausius equation. The thermodynamic function he introduced, known as the Helmholtz free energy, serves as a valuable tool for assessing the direction and limitations of physical and chemical processes under conditions of constant temperature and constant volume. In 1847, he delivered a notable lecture on the conservation of force to the German Physical Society, which significantly enhanced his reputation within the scientific community. During this presentation, Helmholtz was the first to articulate the law of energy conservation using a mathematical framework. Throughout his career, Helmholtz conducted extensive research and made substantial contributions not only to physics but also to physiological optics, acoustics, mathematics, and philosophy.

Lord Kelvin (1824–1907), a prominent English physicist and inventor, made significant contributions to the study of capillary phenomena and their implications. From a young age, Kelvin exhibited remarkable intelligence and a strong dedication to his studies, entering the preparatory school of the University of Glasgow at the age of 10. In 1851, he was elected as a member of the Royal Society of London and subsequently served as its president from 1890 to 1895. Additionally, he was elected as a fellow of the French Academy of Sciences in 1877. Kelvin conducted extensive research and made significant contributions to various fields, including thermology, electromagnetics, fluid mechanics, optics, physical geography, mathematics, and engineering applications. During his time, he garnered a distinguished reputation within the scientific community and received high praise from scientists and academic institutions in Great Britain and other Western countries. His research in thermology and electromagnetics, along with their engineering applications, was particularly noteworthy. In the area of capillarity, he analyzed the differing saturation vapor pressures between liquids and droplets on flat surfaces, leading to the derivation of the Kelvin equation through the application of energy principles. The Kelvin equation serves to elucidate phenomena such as capillary condensation, artificial rainfall, and the radius of the meniscus.

Albert Einstein (1879–1955), the American physicist, is widely regarded as the pioneer and founding figure of modern physics. He is best known for his formulation of the theory of relativity and the mass–energy equivalence principle. Notably, Einstein's initial thesis focused on capillary mechanics. He completed his first thesis, titled *Obtained from the Capillary*, in December 1900, which was subsequently published in the *Physics Journal Letters* the following year. In this work, Einstein derived a law of action pertaining to intermolecular forces as they relate to capillarity.

Despite the extensive historical development of capillary mechanics and its theoretical frameworks that inform practical applications, it has yet to establish itself as an independent discipline. Currently, it remains a subset or branch of thermodynamics, fluid mechanics, or surface science. However, advancements in M/NEMS technology, improvements in micro-nano material science, and the ongoing miniaturization of micro- and nano-mechanical structures have broadened the scope of applications, prompting deeper exploration of fundamental issues. Through this exploration, researchers have identified the significant influence of capillary mechanics on micro- and nano-mechanical structures. Consequently, as a critical surface force, capillary mechanics has garnered increased attention. Nevertheless, a comprehensive understanding of many essential questions within capillary mechanics remains elusive, and numerous unresolved issues persist. Therefore, it is imperative for capillary mechanics to evolve into an independent discipline, integrating with the advancements in micro-nano technology and science, to establish a modern research field that will enhance the study of capillarity.

Chapter 2

Basic Conceptions

2.1 State of Matter and Phase

2.1.1 *State of matter*

The diverse substances present in the natural world are composed of numerous microscopic particles, including molecules, ions, and atoms. These particles engage in random thermal motion while simultaneously interacting with one another. The attractive forces among the particles promote aggregation, whereas thermal motion tends to facilitate their dispersion. At specific temperature and pressure conditions, these opposing effects reach a state of equilibrium, resulting in the clustering of particles into a stable structural configuration, commonly referred to as a state of matter. From a macroscopic perspective, the most prevalent states of matter on Earth are solid, liquid, and gas. In the solid state, matter retains a fixed volume and shape. In the liquid state, matter maintains a constant volume but conforms to the shape of its container. In the gaseous state, matter expands to fill any available volume.

On a microscopic level, solid materials are composed of particles that are arranged in a regular and symmetrical order, a configuration referred to as the crystalline state. The particles that constitute these crystalline structures are interconnected by various types of chemical bonds, including covalent, ionic, and metallic bonds. Consequently, there exists a strong interaction force among the particles. These bonded particles exhibit random vibrations around their equilibrium positions but are unable to move away from them. As a result, solids possess a stable shape and a definite volume. In contrast, while liquids also exhibit strong interaction forces among their constituent particles and their molecular arrangement is often similar to that of solids, this arrangement is typically confined to very small regions (on the order of nanometers). However, the majority of the liquid's volume consists of regions that are completely unordered. Liquid particles vibrate randomly around their equilibrium positions for brief moments before relocating to new equilibrium positions, which allows liquids to flow. The duration of these vibrations varies among different liquids and under different environmental conditions; however, at a given pressure and temperature, the average duration of particle vibration is constant, a phenomenon known as residence time. For instance, the residence time of liquid metal molecules is on the order of 10^{-10} seconds, while that of water molecules is approximately 10^{-11} seconds. For a specific liquid, an increase in temperature results in a decrease in residence time, with shorter residence times correlating to enhanced liquidity. The microscopic characteristics of gases differ significantly from those of solids and liquids. The typical distance between neighboring gas molecules is considerably larger, resulting in diminished intermolecular forces. The motion of gas molecules can be approximated as uniform linear motion until they collide with other molecules or the walls of their container. Consequently, gases expand to occupy the entirety of their container. Unlike two immiscible liquids, two different gases can mix uniformly without forming a distinct interface.

2.1.2   *Phase*

A "phase" is defined as a region within a system that exhibits uniform chemical and physical properties. It represents a state of matter, distinct from the term physical state. The concept of phase emphasizes the uniformity of physical and chemical properties, and is not restricted to a single substance. Typically, there are three states of aggregation — liquid, solid, and gas — associated with a given substance. Within each state of aggregation, the chemical and physical properties are consistent throughout any portion of a pure substance, thereby categorizing each state of aggregation as a phase. Different states of aggregation correspond to different phases; for instance, a system exhibiting two states of aggregation will be characterized by two distinct phases. If a system exhibits two aggregation states, it can be characterized as having two distinct phases. For instance, at a pressure of 101.325 kPa and a temperature of 373.15 K, water and vapor achieve equilibrium and coexist within the system. Consequently, the system comprises two phases: the liquid phase and the gas phase of water. An interface exists between these phases, which possess differing chemical and physical properties. As a result, the two phases can be separated using physical methods. At the macroscopic level, the transition in the nature of the interface is abrupt and discontinuous, allowing for clear differentiation between the phases.

In the three states of matter, gas molecules are characterized by their ability to mix uniformly, resulting in a single phase regardless of the number of substances present in the system. In contrast, liquids exhibit different behavior. When two types of liquids can mix uniformly, they create a homogeneous phase. Conversely, if they cannot mix uniformly, they will separate into two distinct phases, forming an interface. The solid state is more complex than both gas and liquid states. Typically, solids composed of different substances are classified as belonging to different phases.

Under varying temperature and pressure conditions, phases can undergo transformations into one another. This phenomenon is referred to as phase conversion, a prevalent physical process observed in nature. Phase conversions can be categorized into first-order and second-order conversions. During first-order phase transitions, significant changes in volume occur, accompanied by the absorption or release of latent heat. A common example of this phenomenon is the transformation of water from liquid to vapor or ice under specific conditions. In contrast, second-order phase transitions involve changes in the system that do not exhibit the aforementioned characteristics; that is, there are no significant volume changes, nor is there any absorption or release of latent heat. However, during these transitions, certain properties, such as the heat capacity and the coefficient of thermal expansion, may undergo abrupt changes. For instance, some materials that exhibit electrical resistance, including insulators, at room temperature may experience a sudden loss of resistance when the temperature is lowered to a critical threshold. This transition from a normal phase to a superconducting phase exemplifies a second-order phase transition.

2.2   Surface, Interface, and Bulk

2.2.1   *The connotation of surface, interface and bulk*

"Surface" usually refers to the border between the gas phase and the condensed phase (liquid phase and solid phase), including the border between gas and liquid and the one between gas and solid. "Interface" is the border between two condensed phases, including the border between liquid and solid, the one between two kinds of liquid, and the one between two kinds of solid. From the point of view of phase, the concept of interface is more common and more general than surface. The surface of liquid–gas or solid–gas can be considered as the interface between the gas phase and the condensed phase. Therefore, a surface is also an interface between phases. "Bulk" (body) is a term which is different from surface or interface. In order to distinguish it from a surface or an interface, the non-interface internal part of a state of matter or a phase (or surface) is called bulk. Generally, bulk indicates the same kind of non-interface phase. The physical and chemical properties are completely uniform and consistent in one kind of bulk.

2.2.2   *Features of interface*

From a macroscopic perspective, an interface is defined as a boundary that separates two distinct phases and is considered to have no thickness. However, from a microscopic perspective, the existence of thickness is acknowledged. Regardless of whether one examines the interface at a macroscopic or microscopic level, it possesses numerous inherent characteristics.

2.2.2.1   *Energetic properties of interface*

When two phases come into contact, a contact region is established. Within this region, the intrinsic properties of the entire system transition from one phase to another. For stability to be achieved, the interface must possess an interfacial free energy. However, when considered over an extended period, this stability is relative, as absolute stability does not exist. Consequently, the stability of the system is confined to a finite duration. According to the principle of least energy, interfacial free energy consistently seeks minimization. In a two-phase system, if the presence of the interface results in elevated free energy, the interface will spontaneously evolve toward a state of minimization. Similarly, the two phases will tend to achieve an extreme degree of separation.

2.2.2.2 *Transition property of surface*

According to the definitions of states of matter and phases, solids, liquids, and gases (referred to as solid phase, liquid phase, and gas phase, respectively) exhibit distinct characteristics that are often manifested at their surfaces and interfaces. Generally, there is no interface between different types of gases, as gases tend to disperse spontaneously and achieve a uniform distribution. Similarly, there is typically no interface between two different liquids.

Therefore, the term "interface" means the one that lies between two kinds of solids, solid and liquid, and two kinds of immiscible liquids, while the term "surface" refers to the one between solid and gas or between liquid and gas. The dynamic behavior of surfaces or interfaces is frequently neglected in discussions concerning the stability of interfaces. In reality, for both gases and liquids, the surfaces that separate internal bodies, as well as those between liquid and gas phases, are characterized by constant and rapid molecular exchange. The exchange of molecules at the surface between two phases (such as liquid and gas or solid and gas) is referred to as surface migration or surface transition. Analyses indicate that at 25°C, the average dwell time of a water molecule at the interface between gas and water is less than 3 ms, while the dwell time of a mercury atom is approximately 5 ms. In contrast, the dwell time of a solid metal atom at room temperature is nearly 10 s. This significant difference in transition properties between the surfaces of solids and liquids results in distinct characteristics for each. A newly formed liquid surface can quickly achieve thermal equilibrium; however, most solid surfaces require a considerably longer duration to reach a complete steady state of thermal stability. It is also noteworthy that certain solids can sublimate directly into the gas phase. Under conditions of dynamic thermodynamic equilibrium, the number of vapor molecules emitted from a liquid (or solid) during evaporation (or sublimation) is equal to the number of molecules returning to the liquid (or solid). At this point, the vapor is referred to as saturated vapor.

2.2.2.3 *Characterization of molecular in the interface area*

Between two distinct phases, there exists a region characterized by asymmetric forces acting on the molecules, which results in an increase in interfacial free energy. When two phases come into contact, a region emerges where molecular size alterations can occur. Within this region, the system experiences a phase transition, shifting from one phase to another. In the case of a non-volatile solid interacting with an inert gas, the interface transition region is approximately the thickness of a single molecule. Consequently, this leads to a sharp transition from the perspective of boundary dynamics, wherein solid molecules abruptly transform into gas molecules at the interface. Conversely, this transition is not as abrupt when a pure liquid interacts with its corresponding gas phase. In this scenario, the transition region spans several molecular units, and the molecular concentration transitions gradually from liquid density to gas density across various density units. The subsequent chapters will provide a more detailed analysis of the interfacial layer.

2.3   Surface Free Energy

The concepts of surface, interface, and bulk (or body) have been delineated in the preceding section. The surface is a specific type of interface, representing the boundary between tangible objects in the condensed phase and the intangible gas in the gas phase. The term "bulk" relates to the interface as well. Both surface and interface possess inherent energy characteristics.

To effectively comprehend the concept of surface free energy, this chapter initially presents the definitions of new surface and free energy. When the body of an object is divided into two parts along an interface in a vacuum or air, two new surfaces are created. These surfaces are referred to as new surfaces.

The concept of free energy originates from thermodynamic theory. From a thermodynamic perspective, when the temperature exceeds absolute zero, the molecules of any substance exhibit thermal motion, which is inherently associated with the energy of the system. Consequently, all objects possess energy, specifically thermal energy. The total thermal energy of an object comprises two components: one is the observable thermal energy, commonly referred to as free thermal energy, and the other is potential thermal energy. Observable thermal energy can be measured using a thermometer, whereas potential thermal energy remains unobservable and reflects chemical effects. The notion of free energy is derived from the concept of free heat. In 1882, the German physicist and biologist Hermann von Helmholtz first introduced the term and defined what is now known as Helmholtz free energy.

The expression is represented as $A = U - TS$, where A is Helmholtz free energy and the first letter of *Arbeit* in German. U is internal energy, T is the thermodynamic temperature of the system, and S is entropy.

Entropy is a thermodynamic property that serves as an indicator of the energy available for performing useful work during a thermodynamic process. In the field of thermodynamics, the concept of entropy is articulated through the second law of thermodynamics. Entropy is recognized as a state function of a material system, which exemplifies the irreversibility of thermal processes and is typically denoted by the symbol S. In an isolated system, the changes that occur invariably lead to an increase in entropy. From the perspective of molecular thermal motion, the molecules within material systems tend to transition from a state of order to one of chaos, primarily due to thermal agitation. The increase in entropy signifies a rise in the degree of molecular disorder. Consequently, entropy serves as a quantitative measure of the disorder or randomness present within a thermodynamic system. It is well established that entropy reflects the level of molecular disorder associated with thermal motion within the system. The term TS represents the portion of internal energy that is incapable of performing useful work, often referred to as irreversible work. Therefore, under specific conditions, free energy serves as an indicator of the degree of freedom or the amount of internal energy available for work within the system. In addition to Helmholtz free energy, there exists another form of free energy known as Gibbs free energy. Its expression is represented as $G = H - TS$, where G is Gibbs free energy. H is the enthalpy of the system. $H = U + PV$, where P is pressure and V is volume. The two types of free energy mentioned above are typically applicable in different contexts. Gibbs free energy is predominantly utilized in isobaric and isothermal processes, while Helmholtz free energy is primarily employed in isochoric and isothermal processes. Generally, free energy is associated with the capacity to perform useful work. Notably, in the field of physics, the term refers to Helmholtz free energy, whereas in chemistry, it commonly denotes Gibbs free energy.

According to the definition of Helmholtz free energy, the total differential is

$$dA = dU - TdS - SdT \qquad (2.1)$$

According to the first law of thermodynamics, in a closed system,

$$dU = TdS - pdV \qquad (2.2)$$

Substituting equation (2.2) into (2.1) leads to

$$dA = -pdV - SdT \qquad (2.3)$$

According to the definition of Gibbs free energy, the total differential is

$$dG = dU + pdV + Vdp - TdS - SdT \qquad (2.4)$$

From the first law of thermodynamics, in a closed system, it is known that

$$dU = TdS - pdV \qquad (2.5)$$

Substituting equation (2.5) into (2.4) leads to

$$dG = Vdp - SdT \qquad (2.6)$$

The bulk (or body) of an object, which represents a specific state of matter or phase, experiences intermolecular attractions among its constituent units, such as molecules or atoms. However, these attractions result in an overall state of equilibrium. In the context of thermodynamics, free energy is present within the system containing the substance. When the bulk is divided along an interface to create two new surfaces, the units situated at the interface become unbalanced due to their interactions. This results in a net attraction for individual units. For one of the newly formed surfaces, the side that remains in contact with the bulk continues to experience cohesive attraction, while the opposite side is either not attracted (in a vacuum) or experiences weak attraction from the surrounding gas (in air). To address this imbalance and achieve a new equilibrium, a tightening effect emerges on the two new surfaces. From a force perspective, the original cohesive attraction, which acts perpendicular to the direction of the new surface, is disrupted. Consequently, a tension force parallel to the direction of the new surface is generated to establish a new balance. From an energy standpoint, this tightening effect indicates that additional free energy (or new free energy) is created on the two new surfaces compared to the free energy of the original system. The free energy associated with the formation of the new surfaces is referred to as surface free energy. For analytical convenience, the concept of specific free energy is frequently employed, which quantifies the amount of surface free energy per unit area.

The preceding analysis and description regarding the generation and implications of surface free energy are primarily qualitative in nature. This qualitative understanding is instrumental in elucidating the concept and formation mechanisms of surface free energy. However, in practical applications, a quantitative description of surface free energy proves to be more beneficial. From a quantitative perspective, surface free energy is defined as the reversible work required to separate one interface from the bulk material. The prior analysis indicates that individual units within the bulk are subject to attractive forces from neighboring units, resulting in a cohesive effect. When a unit located at an interface is separated along this interface, the cohesive forces must be overcome, which is accomplished through the work performed by an external force. The work performed to counteract the cohesive forces is subsequently stored at the interface as potential energy, which is referred to as surface free energy. An analysis of intermolecular forces indicates that this attractive cohesive force diminishes at a rate approximately proportional to the sixth power of the distance for a single unit (molecule or ion). In contrast, for larger objects, such as semi-infinite bodies, this attractive cohesive force decreases at a rate ranging from quadratic to quartic power with respect to the distance. Generally, this force is characterized as a short-range interaction. When the separation distance falls within the sub-micron to micron range, the force is nearly completely attenuated. Consequently, the phenomenon commonly referred to as interface separation is typically considered to occur within the sub-micron to micron scale. The characterization of magnitude in relation to separation distance is relatively straightforward. In contrast, the quantitative characterization of cohesive attraction presents a greater degree of complexity. Consequently, the method for quantitatively describing surface free energy is inherently challenging. From this perspective, the force is considered to be perpendicular to the surface; thus, analyses typically adopt this viewpoint when discussing adhesive force, adhesive strength, or cohesion. It is noteworthy that free energy is seldom analyzed from this angle. Instead, the trend in tension and the additional (new or surplus) free energy associated with the newly created surface are employed to describe surface free energy. When one phase is in the gas state, its interface is defined as a surface. Therefore, surface free energy can be articulated in terms of a tensioning or tightening force that is parallel to the surface and acts upon it. In light of the observed trend in tension and the additional energy associated with the newly formed surface, it is necessary to apply an external force to separate the interface. The work performed by this external force is equivalent to the work required to overcome the original cohesive forces at the interface. When surface tension is employed to characterize the tension level of the new surface, it follows that, from the perspective of overcoming surface tension, the surface free energy must correspond to the work expended in the process of separating (or forming) the new surface. The fundamental principle underlying this analysis remains consistent with the previous discussion. The energy exerted by the bulk material on the interface manifests as a tightening tendency at the interface. Consequently, a suitable definition can be articulated as follows: specific surface free energy (commonly referred to as surface free energy) is defined as the work required to overcome surface tension (or the tightening force) per unit area.

2.4   The Interface Free Energy and Interfacial Tension

The aforementioned surface represents the interface between the condensed phase and the gas phase. The energy primarily pertains to the interfacial free energy associated with both the condensed phase and the gas phase. A comparable additional free energy is also present at the interface between two condensed phases. The function of interfacial free energy is to stabilize the interface, which results in the minimization of interfacial free energy and maximizes the separation between the two phases.

When two distinct condensed phases come into contact, an interface is established. If both phases are liquid, as in the case of two incompatible pure liquid phases, a liquid–liquid interface is formed. Conversely, when one phase is solid and the other is liquid, a liquid–solid interface is created. In instances where two different solid phases come into contact, a solid–solid interface is generated. However, due to the surface characteristics of solids, achieving close contact between solid–solid interfaces to attain thermal equilibrium is often challenging. Consequently, the interfaces that are typically discussed are the solid–liquid interface and the liquid–liquid interface, which arises from the interaction of two immiscible liquids. Given that the physical states on either side of the interface differ, the cohesive interactions among the material molecules on each side are also distinct. Following a similar principle of surface free energy, a new interfacial free energy is generated. When one of the phases is gaseous, the interfacial free energy reduces to surface free energy. In cases where two identical liquid phases come into contact, the asymmetry is eliminated, resulting in an interfacial free energy of zero. Thus, the interfacial free energy serves as an indicator of the asymmetric attraction present on both sides of the interface. Additionally, there are scenarios in which the interfacial free energy is zero despite the two phases being different; in such cases, the contacting phases do not form an interface. This phenomenon can occur, for example, when a non-wetting liquid contacts its corresponding solid or when two different compatible liquids interact.

The phenomenon of interfacial tension can be likened to surface tension, as it represents a force that acts to contract the interface. This force arises from the asymmetric attraction exerted by the two phases on either side of the interface, resulting in a tightening effect.

2.5 Cohesion and Adhesion

The pervasive existence of intermolecular forces results in the integration of substances into various forms, commonly referred to as the states of matter. Molecules of the same type tend to aggregate, while molecules of different types can also form aggregates. The attractive forces between identical molecules, which lead to the cohesion of similar materials, are termed cohesion forces; these forces operate at the material level. Conversely, the bonding effect that occurs between two different materials due to the attractive forces between dissimilar molecules is known as adhesion forces, which also function at the material dimension. Adhesion effects can be categorized based on various mechanisms, including mechanical adhesion, chemical adhesion, dispersion adhesion, electrostatic adhesion, and diffusion adhesion. Mechanical adhesion, akin to the sewing effect, occurs when the adhesive material penetrates the pores or voids at the interface. In contrast, chemical adhesion involves the formation of compounds at the junction of materials, resulting in adhesion through ionic bonds, covalent bonds, and other types of chemical interactions. Dispersion adhesive can be characterized as a form of adsorption, which pertains to the adhesion of two materials through van der Waals forces. Electrostatic adhesion, on the other hand, involves conductive materials that acquire a charge, resulting in a charge differential at their interface. Consequently, this interaction creates a structure analogous to a capacitor, wherein electrostatic attraction occurs between the two plates of the capacitor-like configuration. When the molecules of the two materials exhibit high levels of activity, they may undergo mutual dissolution and infiltration, leading to the formation of a polymer. This phenomenon is referred to as diffusion adhesion.

A phenomenon can be attributed to a multitude of forces, exemplified by a drop of water falling onto the surface of a leaf. The cohesive forces contribute to the formation of a water droplet, while surface tension causes the droplet to assume a nearly spherical shape. Additionally, adhesive forces facilitate the attachment of the droplet to the leaf's surface.

The concepts of cohesion and adhesion are inherently complex; therefore, the terms cohesion work and adhesion work are frequently employed to elucidate these phenomena.

Cohesive work, also referred to as cohesive energy, is defined as the reversible work required to separate a unit area of material into two distinct surfaces. This concept can be readily understood through the principles of free surface energy. Consequently, cohesive work is indicative of the inseparability resulting from the gravitational forces acting between the internal molecules of the material. When an interface is separated, it results in the formation of two new surfaces. Therefore, the work of adhesion is expected to be twice the value of the surface free energy. If the surface free energy of the material is γ, the adhesion work (energy) is $W_c = 2\gamma$.

The adhesion work, also referred to as adhesion energy, is defined as the reversible work required to separate two distinct materials per unit area along their interface. This adhesion work is not solely dependent on the surface free energy of the individual materials; it is also influenced by the interfacial free energy that exists between them. This concept underscores the inseparability of the two materials at the interface. When two different materials come into contact, a new interfacial free energy is generated at the interface. Upon separation, the total free energy of the surfaces of the two newly formed materials is the sum of their respective surface free energies. However, the interfacial free energy dissipates upon separation. Consequently, the net increase in energy of the system following separation is equivalent to the difference between the combined surface free energies of the two new surfaces and the original interfacial free energy. If the surface free energy of material 1 is γ_1, the surface free energy of material 2 is γ_2, and the interfacial free energy between them is γ_{12}, then the adhesion work (energy) is $W_a = \gamma_1 + \gamma_2 - \gamma_{12}$.

2.6 Interface Layer and Surface Force

The interface layer is a thin region that exists at the two-phase boundary, in contact with each of the two bulk phases. Consequently, the interface layer is characterized as three-dimensional. To differentiate it from the three-dimensional interface layer, the two-dimensional interface between the two phases is referred to as the physical interface. The presence of the physical interface signifies that there are variations in various physical quantities, including composition, on either side of the physical interface. Concerning the bulks on both sides of the interface, this alteration can be classified as a mutation, often referred to as a step change. However, it is important to note that this change should not be perceived merely as a simple mutation between the two phases; rather, it should be understood as a continuous process of transformation. This ongoing process is manifested at the interface layer. Given that the physical characteristics and chemical properties of the interface layer are non-uniform, it can be argued that the interface layer does not constitute a singular phase. Current research in interface science identifies three distinct types of interface layer models. The first model is the Gibbs surface segmentation model. The second model is the Guggenheim interface transition model. The final model is the physical interface model. Essentially, the Gibbs surface segmentation model represents a digital interface layer; however, the physical significance of this interface layer remains ambiguous. The Guggenheim interface transition model is a modification of the Gibbs surface segmentation model. This model treats the interfacial transition zone as a thermodynamic entity, even referring to it as an interface phase. Both models mentioned tend to obscure the physical interface between the two phases, leading to the perception that the macro-interface is equivalent to the micro-interface, with each interface consisting of just one interface layer. Consequently, both models exhibit certain limitations. The physical interface type of the interface layer model first establishes the existence of the physical interface. Furthermore, this model posits that there are two thin interface layers that are separately connected to the bulk materials on either side of the physical interface. These two thin interface layers possess a thickness equivalent to the radius of molecular interaction. Notably, the properties of the two interface layers exhibit a sudden change on both sides of the physical interface, a phenomenon referred to as a step change.

Molecules are influenced by both the interfacial adhesion forces of the material and the cohesive forces at the interface. When the interface layer is considered as a distinct entity, it is essential to analyze its overall force by examining both the internal and external forces acting on the surface.

As previously stated, interactions among identical molecules result in cohesive forces, while interactions between dissimilar molecules give rise to adhesive forces. Within a particular phase of the bulk, cohesive forces are uniformly present throughout. At an interface within the bulk (where the same phase is maintained), cohesive forces manifest as a pulling force that is perpendicular to the interface, a phenomenon that can be referred to as negative pressure. The negative pressure exerted by the same kind of molecules per unit area is referred to as molecular internal pressure. At the interfaces between two different phases, the distinct types of molecules influence the interface, resulting in an adhesive pulling force (negative pressure) that is perpendicular to the interface, which can be termed molecular external pressure. On both sides of the physical interface layer, one side is connected to the bulk phase, while the other side is connected to another interface layer. The side connected to the bulk phase is influenced by molecular internal pressure, whereas the side connected to the interface phase is affected by molecular external pressure.

The integration of molecular internal pressure across the thickness of the interface layer is defined as the surface internal force. Similarly, the integration of molecular external pressure along the thickness of the interface layer is defined as the surface external force. After performing this integration and neglecting the thickness of the interface layer, we can analyze the cohesive effects of the bulk material and the adhesive effects of the adjacent phase at the interface. Consequently, by disregarding the thickness of the interface layer, the surface internal force represents the attractive cohesion of the liquid bulk per unit area of the interface layer. The surface external pressure refers to the attractive adhesion of the adjacent phase (the interface layer of the other phase) per unit area of the interface. For an interface layer with negligible thickness, one side is influenced by surface internal forces, while the other side is influenced by external surface forces. The dimensions of both surface internal and external forces are equivalent to the dimension of surface tension, as they all represent force per unit length. However, it is important to note that both surface internal and external forces act perpendicular to the interface, which distinguishes them from the direction of surface tension. The surface internal force acts toward the interior of the body, while the surface external force is directed toward the two-phase physical interface. Further analysis using molecular theory indicates that the internal surface force is equivalent to half of the cohesive work and is the same as the surface tension. That is,

$$\sigma_{11} = \frac{W_{11}}{2} = \gamma_{11} \tag{2.7}$$

The surface external force is half of the adhesive work and is equal to half the value of the difference between the sum of the two-phase surface internal forces and the interfacial tension:

$$\sigma_{12} = \frac{W_{12}}{2} = \frac{\gamma_{11} + \gamma_{22} - \gamma_{12}}{2} = \frac{\sigma_{11} + \sigma_{22} - \gamma_{12}}{2} \tag{2.8}$$

where σ_{ij} is the surface force. When $i = j$, it is the surface internal force; when $i \neq j$, it is the surface external force. W_{ij} is the surface work. When $i = j$, it is the cohesive work; when $i \neq j$, it is the adhesive work. γ_{ij} is surface (or interface) tension. When $i = j$, it is the surface tension; when $i \neq j$, it is the interface tension.

2.7 Chemical Potential

2.7.1 *The concept of chemical potential*

In a simple closed system — defined as a specific amount of pure substance or a mixture whose composition remains unchanged (with no phase transitions or chemical changes) — only energy (in the form of heat and work) is exchanged with the environment, without any material exchange. Consequently, two thermodynamic variables, such as temperature and pressure, can be utilized to determine the state of the system. However, in open systems or systems with variable composition, this approach is no longer applicable. Under the condition of constant temperature and pressure ($dT = 0$ and $dp = 0$), Gibbs free energy, described by the formula $dG = -SdT + Vdp$, is zero ($dG = 0$). The Gibbs free energy of the system remains unchanged. This statement is clearly incorrect for an open system, as the addition of substances to the system will result in an increase in the system's Gibbs free energy. For instance, when reactive substances are added to a battery, the electrical power produced by the battery must increase. This increase is a direct consequence of the rise in Gibbs free energy within the system. Similarly, under conditions of constant temperature and pressure, if a spontaneous chemical change occurs in a closed system, $dG_{T,p}$ should be less than zero. This is due to the fact that an irreversible change in the composition of the system has taken place.

However, according to formula (2.5), since $dT = 0$ and $dp = 0$, $dG_{T,p} = 0$, which is clearly inconsistent with the practical situation. Therefore, equation (2.5) is not applicable to the process of changing components in a closed system. These two examples illustrate a common characteristic: when the amount of one or more components changes in a system, this alteration will result in a change in free energy. Therefore, for both open systems and closed systems with varying components, it is essential to not only describe free energy using two thermodynamic variables but also to introduce the factor that causes the change in the thermodynamic function due to the alteration in system composition. Thus, for open systems or closed systems with changing components, G (Gibbs free energy) not only depends on the temperature and pressure but also depends on the amount of each component in the system $n_1, n_2, \ldots$, i.e.,

$$G = f(T, p, n_1, n_2, \ldots) \tag{2.9}$$

The total differential of G can be written as

$$dG = \left(\frac{\partial G}{\partial T}\right)_{p,n_1,n_2,\ldots} dT + \left(\frac{\partial G}{\partial p}\right)_{T,n_1,n_2,\ldots} dp + \left(\frac{\partial G}{\partial n_1}\right)_{T,p,n_2,\ldots} dn_1$$

$$+ \left(\frac{\partial G}{\partial n_2}\right)_{T,p,n_1,\ldots} dn_2 + \cdots \tag{2.10}$$

The above equation shows that the change in the Gibbs free energy of the system is equal to the sum of the changes in G caused separately by temperature changes dT (when the pressure and the composition remain unchanged), pressure changes dp (when the temperature and the composition remain unchanged), and changes in each component $dn_1, dn_2, \ldots$ (when the pressure, temperature, and other components remain unchanged in addition to a particular component). The partial derivatives $(\partial G/\partial T)_{p,n_1,n_2,\ldots}$ and $(\partial G/\partial p)_{T,n_1,n_2,\ldots}$ in the above equation are the same as $(\partial G/\partial p)_T$ and $(\partial G/\partial T)_p$ of the closed system, respectively. The rates of change in Gibbs free energy resulting from variations in temperature or pressure, while maintaining a constant composition, are all equivalent. Consequently, the rate of change corresponds to the relationship of Gibbs free energy in a closed system where the composition remains unchanged. That is,

$$\left(\frac{\partial G}{\partial T}\right)_{p,n} = -S \tag{2.11}$$

$$\left(\frac{\partial G}{\partial p}\right)_{T,n} = V \tag{2.12}$$

Then,

$$dG = -SdT + Vdp + \left(\frac{\partial G}{\partial n_1}\right)_{T,P,n_2,\ldots} dn_1$$

$$+ \left(\frac{\partial G}{\partial n_2}\right)_{T,P,n_1,\ldots} dn_2 + \cdots \tag{2.13}$$

To simplify formula (2.13), in 1875, Gibbs proposed using the chemical potential to replace the partial derivative $(\partial G/\partial n_i)_{T,p,n_{j\neq i}}$, i.e.,

$$\mu_i(T, p, n_1, n_2, \ldots) = \left(\frac{\partial G}{\partial n_i}\right)_{T,p,n_{j\neq i}} \tag{2.14}$$

In this function, $n_{j\neq i}$ means that except the component n_i, others components remain unchanged. The chemical potential, as described by the equation above, is a function of temperature, pressure, and the composition of the system. Therefore, the physical significance of chemical potential is the change in Gibbs free energy (often referred to as the Gibbs free energy change rate) that results from the addition of a unit of component *I*. This addition is so small that it does not significantly alter the composition of the system under constant temperature and pressure. The formula can be written as

$$\mu_i(T, p, n_1, n_2, \ldots) = \lim_{\Delta n_i \to 0}\left(\frac{\Delta G}{\Delta n_i}\right)_{T,p,n_{j\neq i}} = \left(\frac{\partial G}{\partial n_i}\right)_{T,p,n_{j\neq i}} \tag{2.15}$$

The concept of chemical potential can also be understood as the change in Gibbs free energy that would occur when 1 mole of a substance is added to a large system. Since the system is very large, the addition of 1 mole of material will not significantly alter the composition of the system. A single-component system only contains one component. Therefore, (2.14) can be written as

$$\mu = \left(\frac{\partial G}{\partial n}\right)_{T,p} = G_m \tag{2.16}$$

The aforementioned equation indicates that, in a single-component system, the chemical potential is equivalent to the rate of change of Gibbs free energy when the substance is in a pure state. This concept is also referred to as molar Gibbs free energy.

The definition of chemical potential indicates that it is a physical parameter related to the strength properties of a system and serves as a state function. Under conditions of constant temperature and pressure, the value of chemical potential varies solely with changes in composition and is independent of the total amount of the system. After introducing chemical potential μ_i, (2.13) can be written as

$$dG = -SdT + Vdp + \sum_i \mu_i dn_i \tag{2.17}$$

Similarly, if the internal energy depends on the volume, entropy, and composition of the system, then

$$U = f(S, V, n_1, n_2, \ldots) \tag{2.18}$$

The enthalpy H is the composition of the pressure and entropy of the system. Helmholtz free energy A is decided by the temperature and the volume. That is,

$$H = f(S, p, n_1, n_2, \ldots) \tag{2.19}$$

$$A = f(T, V, n_1, n_2, \ldots) \tag{2.20}$$

Then, using the same method as above, it is not difficult to obtain the relationships about the chemical potential:

$$\mu_i = \left(\frac{\partial U}{\partial n_i}\right)_{S,V,n_{j\neq i}} = \left(\frac{\partial H}{\partial n_i}\right)_{S,p,n_{j\neq i}} = \left(\frac{\partial A}{\partial n_i}\right)_{T,V,n_{j\neq i}}$$

$$= \left(\frac{\partial G}{\partial n_i}\right)_{T,p,n_{j\neq i}} \tag{2.21}$$

Correspondingly, the total differential of its internal energy U, enthalpy H, and Helmholtz (A) free energy can be written as

$$dU = TdS - pdV + \sum_i \mu_i dn_i \tag{2.22}$$

$$dH = TdS + Vdp + \sum_i \mu_i dn_i \tag{2.23}$$

$$dA = -pdV - SdT + \sum_i \mu_i dn_i \qquad (2.24)$$

$$dG = -SdT + Vdp + \sum_i \mu_i dn_i \qquad (2.25)$$

The four equations above are suitable for the basic equation of chemical thermodynamics in open systems and closed systems in which the composition changes. For a process under constant temperature and pressure, we have

$$dG_{T,p} = \sum_i \mu_i dn_i \qquad (2.26)$$

When a reversible change (phase change or chemical change) in the composition occurs or the system reaches equilibrium, the change in Gibbs free energy should be zero. That is,

$$dG_{T,p} = 0, \sum_i \mu_i dn_i = 0 \qquad (2.27)$$

With an irreversible change (phase change or chemical change) in the composition, Gibbs free energy should decrease:

$$dG_{T,p} < 0, \sum_i \mu_i dn_i < 0 \qquad (2.28)$$

Combining the above two equations,

$$\sum_i \mu_i dn_i \leq 0 \qquad (2.29)$$

The equation is applicable to the reversible process, while the inequality is applicable to the irreversible process.

2.7.2 *The relationship between chemical potential and temperature pressure*

The preceding discussions regarding chemical potential predominantly focus on isothermal and isobaric conditions. This emphasis serves to illustrate the relationship between chemical potential and the rate of change of the relative components of Gibbs free energy under these specific conditions. However, this should not be interpreted as an indication that there is no relationship between chemical potential and temperature or pressure. On the contrary, chemical potential is intricately linked to both temperature and pressure.

According to the definition of chemical potential (2.14) and (2.12), the relationship between the chemical potential and pressure is

$$\frac{\partial \mu_i}{\partial p} = \frac{\partial \left(\frac{\partial G}{\partial n_i} \right)}{\partial p} = \frac{\partial \left(\frac{\partial G}{\partial p} \right)}{\partial n_i} = \frac{\partial V}{\partial n_i} \tag{2.30}$$

Since there is only one component, for the single-component system, we have

$$\frac{\partial V}{\partial n_i} = \frac{\partial V}{\partial n} = V_m \tag{2.31}$$

Then,

$$\left(\frac{\partial \mu}{\partial p} \right)_T = V_m \tag{2.32}$$

That is, for pure components, under a constant temperature, the change rate of chemical potential to pressure is equal to the molar volume.

Similarly, from (2.14) and (2.12), the relationship between chemical potential and temperature can be derived as

$$\frac{\partial \mu_i}{\partial T} = \frac{\partial \left(\frac{\partial G}{\partial n_i} \right)}{\partial T} = \frac{\partial \left(\frac{\partial G}{\partial T} \right)}{\partial n_i} = -\frac{\partial S}{\partial n_i} \tag{2.33}$$

For a single-component system, we have

$$\frac{\partial \mu}{\partial T} = -S_m \tag{2.34}$$

That is, for a single-component system, under a constant pressure, the change rate of chemical potential to temperature is equal to a negative molar entropy.

2.7.3 The application of chemical potential in the phase transition

If under a given temperature and pressure, the liquid in a closed container evaporates, the evaporation tends to a certain degree of saturation and then reaches gas–liquid equilibrium.

According to (2.32), the chemical potential can be used to calculate the change in the Gibbs free energy of the system:

$$dG_{T,p} = \sum_i \mu_i dn_i = \mu_l dn_l + \mu_g dn_g \qquad (2.35)$$

Because the number of water liquid disappeared (dn_l) is equal to the number of water vapor appeared ($+dn_g$), i.e., $-dn_l = dn_g$, the above equation can be written as

$$dG_{T,p} = (\mu_g - \mu_l)dn_g \qquad (2.36)$$

When the water irreversibly transfers into vapor, $\mu_g - \mu_l < 0$, i.e., $\mu_g < \mu_l$.

Thus, when the water irreversibly transfers from liquid into vapor, it always travels from the high-chemical-potential liquid phase to the low-chemical-potential gas phase until the water and vapor reach the chemical potential equilibrium. The condition of gas–liquid equilibrium is

$$\mu_{H_2O,l} = \mu_{H_2O,g} \qquad (2.37)$$

Extending the above results to a multi-component equilibrium system, when it does not reach equilibrium, component B in the prior system always travels from the high chemical potential to the low chemical potential until all the chemical potentials are equal. Therefore, the equilibrium condition of each phase of component B in the multiphase system is

$$\mu_B^\alpha = \mu_B^\beta = \mu_B^\gamma = \cdots \qquad (2.38)$$

In the state of saturation, since the transition between gas and liquid phases is reversible, the change in the free energy of the system is zero.

2.7.4   *Ideal gas chemical potential*

The preceding analysis indicates that, under specific conditions — such as constant temperature and pressure, constant entropy and volume, constant pressure, constant entropy, or constant temperature and volume — the criterion for a system to achieve phase equilibrium is that the chemical components across different phases possess identical chemical potential. However, it is important to note that this represents a general principle. A detailed characterization of the chemical potential for particular substances and systems is necessary for a comprehensive understanding.

For a single-component ideal gas, it is known by (2.34) that

$$\frac{\partial \mu}{\partial p} = V_m \tag{2.39}$$

Therefore, change in chemical potential can be written as

$$d\mu = V_m dp \tag{2.40}$$

For an ideal gas, $V_m = RT/p$, where T is absolute temperature and R is the gas constant. Substituting it into (2.40) leads to

$$d\mu = \frac{RT}{p} dp = RT d\ln p \tag{2.41}$$

If the pressure changes from p_1 to p_2, the chemical potential changes into

$$\Delta\mu = RT \ln \frac{p_2}{p_1} \tag{2.42}$$

Chapter 3

Surface Tension and Interfacial Tension

3.1 Conception of Surface Tension

The concept and mechanism of surface free energy were introduced in the previous chapter. Surface free energy is a form of potential energy. There are three primary reasons for the formation of surface free energy. The first is the creation of a new surface. The second is the cohesive forces present in the bulk material. The third is the asymmetry or imbalance produced by these cohesive forces, which results in a tendency for tension on the surface. The dwell time of surface molecules varies across different phases. In the liquid phase, the dwell time is so short that the system quickly reaches thermodynamic equilibrium. Consequently, the surface tension of the liquid phase stabilizes rapidly. In contrast, the transfer rate of solid surface molecules is relatively slow, resulting in a longer dwell time for the solid phase. As a result, achieving thermodynamic equilibrium in the solid phase within a short time frame is challenging. Therefore, the macroscopic characterization of its surface tension trend is not readily apparent. The surface free energy serves as a measure of surface tension; however, it lacks directionality. To provide an initial description of the directional aspect of tension, which characterizes force, the concept of surface tension is defined. Thus, surface tension describes the degree of tension at the surface, encompassing both magnitude and directional characteristics. The tension exists in the two-dimensional tangential direction of a surface, and its magnitude reflects the degree of tension. As a two-dimensional force, it generally indicates the force density, representing the tension per unit length. Therefore, surface tension can be defined as a tension force acting along a unit length that is parallel to the surface and perpendicular to a specific straight line within the surface layer. This definition is rooted in mechanics. The effect of surface tension on solid surfaces is not as pronounced, as the time required for a solid surface to reach thermodynamic equilibrium differs significantly from that of a liquid surface. Consequently, the concept of surface tension is primarily applied to the liquid phase, while free energy is typically used in the context of solid surfaces.

Surface free energy is derived from surface tension, and both concepts are utilized to describe the trends and magnitude of surface tension. Consequently, surface tension can be defined in terms of both mechanics, as previously mentioned, and energy. From an energy perspective, surface tension can be defined as the increase in surface free energy per unit increase in area. In fact, these two definitions are equivalent. For example, when a surface tension γ acts on the surface, in order to overcome the surface tension by an outside force, a work of $\gamma \cdot \delta A$ must be done to increase the surface area δA. This work is stored in the surface in the form of potential energy, which produces the new surface free energy. According to the definition of free energy, if the new surface free energy is divided by the increased area δA, the surface tension in mechanics definition can be easily obtained.

There are several phenomena that can help us understand surface tension from a mechanical perspective. These phenomena can be categorized into two cases: statics and dynamics. A coin cannot float on the surface of water solely due to the buoyant force, as the mass density of the coin is significantly greater than that of water. However, when a coin is gently placed on the water's surface, it does not sink, as illustrated in Figure 3.1, which represents a static phenomenon of surface tension. In Figure 3.2, a water strider can effortlessly move across the water's surface using its small legs, exemplifying a dynamic phenomenon of surface tension. Both the floating coin and the moving water strider are influenced by surface tension.

There are two more experiments explaining surface tension from the mechanical standpoint. One is pulling soapy water by a piece of

Figure 3.1. Coin floats on water.

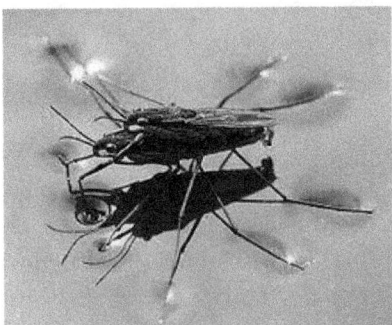

Figure 3.2. Water strider moves on water.

wire, and the other is putting a cylindrical needle floating on water surface. In Figure 3.3, a soap film is formed when a piece of wire on a wire frame is pulled slowly after the frame dipped in soapy water. The phenomenon shown in Figure 3.3 can be described by mechanics as follows.

If the length of the moving wire is l, the stretching force is F, and the displacement of moving wire is Δx, the newly formed area will be $2l\Delta x$. If the surface tension of the upper and lower surfaces is γ (tension on a unit length), the total force is $2\gamma l$, which should equal the stretching force $F \cdot F = 2\gamma l$, and so $\gamma = F/2l$. From this point of view, the mechanical meaning of surface tension during the movement can be understood.

Capillary Mechanics

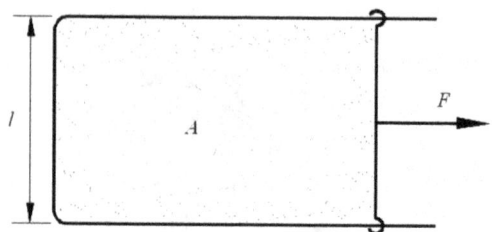

Figure 3.3. A metal frame pulls a soap film.

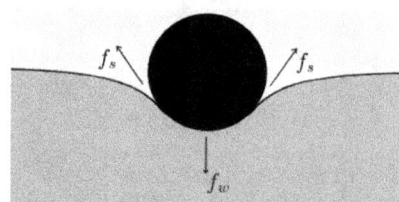

Figure 3.4. A columniform needle presses on water.

The other experiment is shown in Figure 3.4. A cylindrical needle floats on the surface of water. Its weight is f_w. The needle cannot sink because its weight is balanced by the surface tension forces f_s. f_s is in the tangent direction in which the needle touches the water and acts on both sides of the needle. Horizontally, the surface tension effects on the two sides are balanced out: $f_s \sin\theta - f_s \sin\theta = 0$. Vertically, surface tension is the same as the weight of the needle. The length of the cylinder is l. The surface tension of the water is γ. Then, $f_s = \gamma l$. $\gamma = f_s/l$

$$f_w = f_s \cos\theta + f_s \cos\theta = 2f_s \cos\theta$$

From this point of view, the mechanical meaning of a static float object under surface tension can be also understood.

Of course, surface tension can be explained in terms of energy. These two definitions are equivalence relations. In the experiment of pulling soap film mentioned above, the work done by the outside force F is $F\Delta x$. This work is stored as a potential energy in the two surfaces of the soap film. The total area of the surfaces is $2\Delta xl$. This energy forms the surface free energy A: $A = F \cdot \Delta x$. The free energy per unit area (also called the surface free energy ratio) is

$\gamma = A/2\Delta xl = F\Delta x/2\Delta xl = F/2l$. Compared to the mechanical characterization of surface tension above, the surface free energy γ is the same as the surface tension γ mentioned above.

3.2 Molecule Meaning of Surface (Interface) Free Energy–Surface (Interface) Tension

The concept and mechanism of surface free energy were introduced in the previous chapter. Surface free energy is a form of potential energy. There are three primary reasons for the formation of surface free energy. The first is the creation of a new surface. The second is the cohesive forces present in the bulk material. The third is the asymmetry or imbalance produced by these cohesive forces, which results in a tendency for tension on the surface. To counteract this imbalance, a tensioning trend (or tightening) develops on the surface. The fundamental internal factor of the formation of surface free energy is the existence of attractive interaction between molecules in the bulk (body); the fundamental external factor is the formation of a new surface. Surface free energy is defined as "a reversible work done to overcome the cohesion between the molecules inside the bulk (body) in order to separate the body along a certain interface". This work is stored in the form of potential energy in the newly produced surface, which is called surface free energy.

According to the molecule theory, the van de Waals force exists between two molecules. Therefore, there is van der Waals interaction potential between two molecules, and its expression is

$$W_{\mathbf{vdw}} = \frac{C_{\mathbf{vdw}}}{r^6} \tag{3.1}$$

where $C_{\mathbf{vdw}}$ is the coefficient of the interaction potential by van der Waals molecular pairs. $C_{\mathbf{vdw}} = C_{\mathbf{orient}} + C_{\mathbf{ind}} + C_{\mathbf{disp}}$, in which $C_{\mathbf{orient}}$ is the coefficient of oriented interaction potential by Keesom molecular pairs, $C_{\mathbf{ind}}$ is the coefficient of induced interaction potential by Debye molecular pairs, and $C_{\mathbf{disp}}$ is the coefficient of dispersion interaction potential by London molecular pairs. There is a minus in the original formula. For convenience of analysis, take the attractive effect as positive. Therefore, the minus can be taken out of the formula.

The magnitude of the van der Waals force between two molecules is

$$f_{\mathbf{vdw}} = -\frac{\partial W_{\mathbf{vdw}}}{\partial r} = \frac{6C_{\mathbf{vdw}}}{r^7} \tag{3.2}$$

It is along the direction of the connection of the two molecules.

According to Hamaker theory, the total interaction potential of the object system is equal to the sum of the interaction potentials among the molecules, so

$$W_{\mathbf{sum}} = \frac{1}{2} \sum_{i=0}^{N} \sum_{j=0(\neq i)}^{N} W_{\mathbf{vdw}}^{ij}(r_{ij}) \tag{3.3}$$

where $W_{\mathbf{vdw}}^{ij}$ is the interaction potential between molecule i and molecule j, which is not affected by any other molecules; r_{ij} is the distance between molecule i and molecule j; and N is the number of molecules.

According to this theory, the interaction potential between a single molecule and an infinite size of plate with certain thickness is

$$\begin{aligned} W(l) &= 2\pi C_{\mathbf{vdw}}\rho_1 \int_l^{l+h} dz \int_0^\infty \left(\frac{x}{(z^2 + x^2)^3} \right) dx \\ &= \frac{2\pi C_{\mathbf{vdw}}\rho_1}{12} \left(\frac{1}{l^3} - \frac{1}{(l+h)^3} \right) \end{aligned} \tag{3.4}$$

where ρ_1 is the number of molecules per unit volume (molecule density) of the plate with an infinite size and a certain thickness, h is the thickness of plate, and l is the nearest distance between a single molecule and the plate.

If the thickness is infinite, the interaction potential between a single molecule and the semi-infinite body can be described as

$$W(l) = 2\pi C_{\mathbf{vdw}}\rho_1 \int_l^\infty dz \int_0^\infty \left(\frac{x}{(z^2 + x^2)^3} \right) dx = \frac{2\pi C_{\mathbf{vdw}}\rho_1}{12l^3} \tag{3.5}$$

If a unit area is used for the plate with infinite size and certain thickness to replace the single molecule, the interaction potential of two plates with infinite size and certain thickness is

$$W(l) = \frac{2\pi C_{vdw}\rho_1\rho_2}{12} \int_l^{l+h_2} \left[\frac{1}{z^3} - \frac{1}{(z+h_1)^3} \right] dz$$

$$= \frac{2\pi C_{vdw}\rho_1\rho_2}{24} \left[\frac{1}{l^2} - \frac{1}{(l+h_2)^2} - \frac{1}{(l+h_1)^2} + \frac{1}{(l+h_1+h_2)^2} \right]$$

(3.6)

where h_1 and h_2 are the thicknesses of the two plates with infinite size, respectively.

When $h_1 \to \infty$, the interaction potential between a plate with infinite size and certain thickness and a semi-infinite body is

$$W(l) = \frac{2\pi C_{vdw}\rho_1\rho_2}{24} \left[\frac{1}{l^2} - \frac{1}{(l+h_2)^2} \right]$$

(3.7)

When $h_1 \to \infty$ and $h_2 \to \infty$, the interaction potential between the two semi-infinite bodies is

$$W(l) = \frac{2\pi C_{vdw}\rho_1\rho_2}{24} \frac{1}{l^2} = \frac{A_{12}}{12\pi l^2}$$

(3.8)

where $A_{12} = \pi^2 C_{vdw}\rho_1\rho_2$ is the Hamaker constant quantity.

In fact, for an interface, there is still a molecular distance for two contacting semi-infinite bulks. If l represents d_{11}, which is the distance between molecules on the interface surface of two semi-infinite object bulks (bodies) whose materials are the same, the interaction potential in the expression above will change into (absolute value is used here for the convenience of explanation)

$$W_1(d) = \frac{2\pi C_{vdw}\rho_1\rho_1}{24} \frac{1}{d_{11}^2} = \frac{A_{11}}{12\pi d_{11}^2}$$

(3.9)

where $A_{11} = \pi^2 C_{vdw}\rho_1^2$ is the Hamaker constant quantity of object 1. This interaction potential is equal to the cohesion work of the body. In terms of the definition of surface free energy (surface tension), when the bulk (body) is separated along an interface, the work for overcoming the cohesion is equal to the interaction potential. Since two surfaces will be formed after the separation, the potential energy in each surface is equal to half of the interaction potential.

Therefore, the surface free energy (surface tension) of the object is equal to half of the interaction potential, so

$$\gamma_1 = \frac{W_1(d)}{2} = \frac{A_{11}}{24\pi d_{11}^2} \tag{3.10}$$

If l represents d_{12}, which is the distance between molecules on the interface of the two phases, and the materials of the two semi-infinite bodies are different, the interaction potential in equation (3.9) will be

$$W_{12}(d) = \frac{2\pi C_{vdw}\rho_1\rho_2}{24} \frac{1}{d_{12}^2} = \frac{A_{12}}{12\pi d_{12}^2} \tag{3.11}$$

This interaction potential energy is equal to the adhesion work between the two phases. Since the adhesion work has the following relation with surface tension (free energy) and interfacial tension (free energy),

$$W_{12} = \gamma_1 + \gamma_2 - \gamma_{12} \tag{3.12}$$

where γ_{12} is the interfacial tension (free energy), the interfacial tension (free energy) can be described as

$$\gamma_{12} = \gamma_1 + \gamma_2 - W_{12} = \frac{1}{12\pi}\left(\frac{A_{11}}{d_{11}^2} + \frac{A_{22}}{d_{22}^2} - \frac{A_{12}}{d_{12}^2}\right) \tag{3.13}$$

where d_{11} is the distance between molecules in object 1, d_{22} is the distance between molecules in object 2, and d_{12} is the distance between molecules on the interface of object 1 and object 2. A_{11} is the Hamaker constant quantity of object 1, A_{22} is the Hamaker constant quantity of object 2, and A_{12} is the Hamaker constant quantity of the subject between object 1 and object 2.

3.3   Concept of Interfacial Tension

Similarly to the concept of surface free energy, interfacial free energy exists at the interface. Similar to surface tension, there is a force at the interface that creates a tendency for tension, known as interfacial tension. This tension arises from the asymmetric attractive forces on either side of the interface. The mechanics definition of interfacial tension is as follows: interfacial tension is a force per unit line length acting at the interface, which is vertical to this straight line and parallel to the interface. The energy definition of interfacial tension is as follows: it is the increment of interfacial free energy resulting from an increase in per unit area of the interface. When one of the phases is a gas, the interfacial tension will be surface tension.

3.4 Factors Influencing Surface Tension

The surface tension is a property of liquids (and, in general, solids) that reflects their strength. Several factors influence surface tension, including the type of material, its density, and environmental conditions such as temperature and pressure.

3.4.1 *Influence of the type of matter on surface tension*

Surface tension is related to the type of matter because it arises from the net attractive forces generated by the asymmetric cohesive effects of surface molecules. This phenomenon depends on the attractive forces between the molecules and their structural characteristics. For example, a water molecule is a polar molecule, which exhibits strong attractive forces between its molecules. Under normal atmospheric pressure, the surface tension of water can be up to $72.75\,\mathrm{mN \cdot m^{-1}}$ at $20°C$. However, under the same conditions, the surface tension of non-polar hexane molecules is only $18.4\,\mathrm{mN \cdot m^{-1}}$. Hydrargyrum has a huge cohesion force. Therefore, at room temperature, hydrargyrum has the greatest surface tension ($\gamma_{\mathbf{Hg}} = 485\,\mathrm{mN \cdot m^{-1}}$) among all kinds of liquids. Of course, other kinds of molten metals have great surface tension (generally, the data are collected in the high-temperature molten state). For example, the surface tension of molten copper can be up to $879\,\mathrm{mN \cdot m^{-1}}$ at $1100°C$. The surface tensions of some materials against that of air are listed in Table 3.1.

3.4.2 *Influence of interface*

The surface tension of a liquid is typically defined as the value measured when the liquid is in contact with air (including its vapor).

Table 3.1. Surface tension of different materials under different temperatures.

**Surface tension of various liquids in dyn/cm against air
Mixture %'s are by mass dyne/cm is also called mN/m
(milli-Newton per meter) in S.I. units**

Liquid	Temperature (°C)	Surface tension γ(mN/m)
Acetic acid	20	27.6
Acetic acid (40.1%) + Water	30	40.68
Acetic acid (10.0%) + Water	30	54.56
Acetone	20	23.7
Diethyl ether	20	17.0
Ethanol	20	22.27
Ethanol (40%) + Water	25	29.63
Ethanol (11.1%) + Water	25	46.03
Glycerol	20	63
n-Hexane	20	18.4
Hydrochloric acid 17.7 M aqueous solution	20	65.95
Isopropanol	20	21.7
Mercury	15	487
Methanol	20	22.6
n-Octane	20	21.8
Sodium chloride 6.0 M aqueous solution	20	82.55
Sucrose (55%) + Water	20	76.45
Water	0	75.64
Water	25	71.97
Water	50	67.91
Water	100	58.85

Surface tension varies with changes in the properties of the matter when it interacts with the liquid. Antonoff discovered that the interfacial tension between two types of liquids is equal to the difference in their surface tensions when they are mutually saturated, even if their mutual solubility is minimal. That is,

$$\gamma_{1,2} = \gamma_1' - \gamma_2' \tag{3.14}$$

where γ_1' and γ_2' are the surface tensions of the two kinds of liquids which are mutually saturated. The state of mutual saturation is that when liquid 1 is saturated by the vapors of liquid 2, and liquid 2 is saturated by the vapors of liquid 1. This is called the Antonoff law.

Table 3.2. Interfacial tension between organic liquids and water ($mN \cdot m^{-1}$).

Liquid	Surface tension			Interfacial tension		
	Water layer σ_1'	Organic liquid layer σ_2'	Pure organic liquid	Calculate value	Measurement value	Temperature (°C)
Benzene	63.2	28.8	28.4	34.4	34.4	19
Aether	28.1	17.5	17.7	10.6	10.6	18
Chloroform	59.8	26.4	27.2	33.4	33.3	18
Carbon tetrachloride	70.9	43.2	43.4	24.7	27.7	18
Pentanol	26.3	21.5	24.4	4.8	4.8	18
{ 5%Pentanol 95%Benzene	41.4	28.0	26.0	13.4	16.1	17

The interfacial tensions between common organic liquids and water are shown in Table 3.2.

In the interface between a liquid and a gas, surface tension is the sum of the static attractive forces generated by the interactions among liquid molecules. In this context, the attraction between liquid molecules and air molecules can be disregarded. However, at the interface between two different liquids, the molecules of each liquid attract one another, which reduces the static attractive force of each type of liquid. Consequently, the tension at the new interface is lower than that of the greater tension of the two original surfaces. In an extreme case, when the two types of liquids are in the same phase, the asymmetry disappears. This leads to symmetrical attraction, resulting in a static attractive force of zero, as well as an interfacial tension that is also zero.

3.4.3 *Influence of temperature*

Since temperature directly affects the activities of the molecules, it directly affects surface tension. When the temperature rises, the material will expand, the distances between molecules will extend, and the surface tension will decrease. The linear relation between temperature and surface tension is shown in Figure 3.5. When the temperature is rising up to the critical point t_c, the interface between the liquid and the gas will disappear, and the surface tension tends to

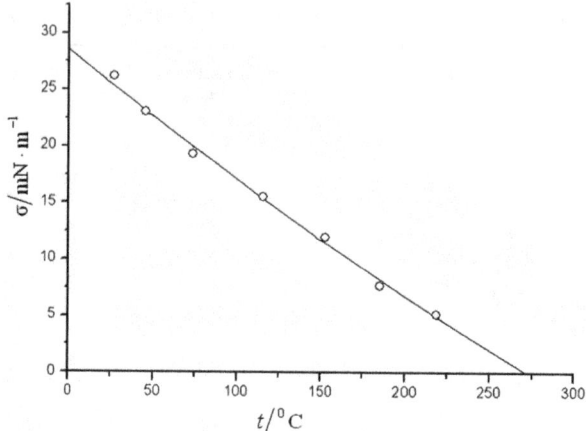

Figure 3.5. $\gamma - t$ curve of CCl_4.

zero. In fact, according to the analysis of the influence from density in the following section, the rising temperature leads to a decrease in the difference of density between the liquid phase and the gas phase. Since the cohesion attraction is proportional to the high power of density, when temperature rises, both the densities and the cohesion attraction of the liquid and gas phases will be close to each other. Then, the surface tension will be zero.

According to their research, many scholars indicate that there is a linear relation between surface tension and temperature. They also think that the rate slope of the linear relation is the changed value of entropy when the surface changes within a unit area. Currently, the descriptions of the relationship between surface tension and temperature are expressed through empirical formulas, the most notable of which is the Eötvös rule.

The Eötvös rule, named after the Hungarian physicist Loránd (Roland) Eötvös (1848–1919), can predict the surface tension of any kind of liquid or substance at any temperature. The prerequisites for this prediction include knowledge of the liquid's density, molar mass, and critical temperature, as well as the understanding that the surface tension approaches zero at the critical point.

Eötvös rule is based on two key assumptions:

1. **Surface tension is a linear function of temperature**. This assumption holds true for most types of liquids. The relationship

is represented by a straight line. At the intersection of this line and the temperature axis, surface tension reaches zero, and the corresponding temperature at this point is known as the critical temperature. Since the critical temperatures of various liquids differ, the Eötvös rule also illustrates the relationship between the surface tensions of different types of liquids.

2. **Data on the relationship between temperature and surface tension** can be represented by a curve, provided that the density, molar mass, and critical temperature of the liquid are known.

Let V be the molar volume and T_c be the critical temperature of a liquid. The surface tension of the liquid is

$$\gamma V^{2/3} = \mathbf{k}(T_c - T) \tag{3.15}$$

where k is the Eötvös constant value for any kind of liquid and has a value of $2.1 \cdot 10^{-7}$ J/(K·mol$^{2/3}$).

Since the intersection of the curve and the temperature axis consistently occurs 6 K (thermodynamic temperature) earlier than the critical point in experiments, Ramsay and Shields modified expression (3.15) to accurately describe the relationship between surface tension and temperature. They achieved this by shifting the expression 6 K to the left along the temperature axis:

$$\gamma V^{2/3} = \mathbf{k}(T_c - T - 6.0) \tag{3.16}$$

The molar volume V can be given by the molar mass M and the density ρ as $V = M/\rho$.

$\gamma_{\mathbf{mol}} = \gamma V^{2/3}$ in the above equation can be called the molar surface tension. Substituting it into the expression above leads to

$$\gamma = \mathbf{k} \left(\frac{M}{\rho} \right)^{-2/3} (T_c - 6.0 - T) \tag{3.17}$$

Introducing the Avogadro constant N_A, the above equation can be rewritten as

$$\gamma = \mathbf{k}' \left(\frac{M}{\rho N_A} \right)^{-2/3} (T_c - 6.0 - T) = \mathbf{k}' \left(\frac{N_A}{V} \right)^{2/3} (T_c - 6.0 - T) \tag{3.18}$$

Table 3.3. Some liquid surface tension values at different temperatures ($mN \cdot m^{-1}$).

Liquid	0°C	20°C	40°C	60°C	80°C	100°C
Water	75.64	72.75	69.56	66.18	62.61	58.85
Ethanol	24.05	22.27	20.60	19.01	—	—
Toluene	30.74	28.43	26.13	23.81	21.53	19.39
Benzene	31.6	28.9	26.3	23.7	21.3	—

As John Lennard-Jones and Corner showed in 1940, by means of statistical mechanics, the constant k' is nearly equal to the Boltzmann constant.

Some of the values of liquid surface tensions at different temperatures are listed in Table 3.3.

3.4.4 *Influence of density*

The difference in densities on either side of the interface results in asymmetric attractions, which contribute to the formation of surface tension or interfacial tension. Consequently, density can influence the magnitude of surface tension. The relation between surface tension and density is

$$\gamma = \mathbf{B}(\rho_l - \rho_0)^4 \qquad (3.19)$$

where ρ_1 is the density of the liquid phase of the liquid and ρ_0 is the density of the gas phase of the liquid. B is a constant, which has nothing to do with the density. This equation is derived from the parachor of interface chemistry.

When the temperature is considered in the relation between surface tension and density, the equation above can be written as

$$\gamma = \mathbf{const}(T_c - T)^p(\rho_l - \rho_0)^q \qquad (3.20)$$

where p and q can be obtained using "least squares" based on experiments. Expression (3.20) is the expansion of expression (3.19).

3.4.5 *Influence of pressure*

The pressure of the gas phase should influence surface tension, given the density difference between gas and liquid, as well as the static attractive forces involved. Since the vapor pressure of a liquid remains constant at a specific temperature, the effect of pressure can only be observed by altering the pressure of air or inert gases. However, it is important to note that air and inert gases can dissolve in and be absorbed by the liquid. Part of the air is attracted to the surface of a liquid. The rate of dissolution and the amount of absorption vary with changes in pressure. Consequently, the method of altering the pressure of air or inert gases to measure surface tension involves multiple influencing factors, including dissolution, absorption, and pressure. Some researchers have investigated the impact of pressure on surface tension. For example, the surface tension of water is $72.82 \, \mathrm{mN \cdot m^{-1}}$ under a pressure of $0.098 \, \mathrm{MPa}$ and $66.43 \, \mathrm{mN \cdot m^{-1}}$ under a pressure of $9.8 \, \mathrm{MPa}$. The surface tension of benzene is $28.85 \, \mathrm{mN \cdot m^{-1}}$ under a pressure of $0.098 \, \mathrm{MPa}$ and $21.58 \, \mathrm{mN \cdot m^{-1}}$ under a pressure of $9.8 \, \mathrm{MPa}$. Therefore, surface tension decreases as pressure increases. However, the effect of pressure on surface tension is not as pronounced as that of temperature. The influence on surface tension is minimal when the pressure changes only slightly.

3.5   Methods of Measurement of Surface and Interfacial Tensions

There are many methods for measuring surface tension, including the capillary rise method, Du Noüy ring method, maximum bubble pressure method, Wilhelmy plate method, spin drop method, droplet volume method, droplet shape analysis method, experimental ink method, metal rod method, and stalagmometeric method.

These methods can measure surface (interface) tension in various environments. The Du Noüy ring method is a classic technique for measuring surface tension, even in situations where wetting is challenging. The principle involves extracting a liquid film (similar to a soap bubble) from the liquid using a ring that is initially immersed in the liquid while measuring the force required to lift the ring. The maximum pressure bubble method is particularly effective for assessing the variation of surface tension over time, as it measures surface tension by determining the highest pressure of the bubbles. The Wilhelmy plate method is a versatile technique, particularly well-suited for the long-term measurement of surface tension. This method quantifies surface tension by assessing the force exerted on a flat plate that is positioned perpendicular to the liquid surface during the wetting process. Additionally, it can be employed to measure interfacial tension. The rotating droplet method is another approach that can be utilized to determine interfacial tension, making it especially effective in scenarios involving low or very low tension. The measured value refers to the diameter of the rotating droplet. The droplet shape analysis method is effective for measuring interfacial tension and surface tension, and it can be applied at very high pressures and temperatures. Surface tension, or interfacial tension, of a liquid can be determined by analyzing the geometric shape of the droplet. The droplet volume method is particularly well-suited for dynamically measuring interfacial tension. The measured value is calculated by dividing the number of droplets by a specific volume of liquid.

Due to the numerous methods available, the following presents only a brief introduction to several commonly used measurement techniques.

3.5.1 *Capillary rising method*

This is one of the simplest methods. When a clean glass capillary tube is placed in a liquid, if the liquid wets the wall of the capillary tube, due to surface tension, the liquid will rise along the capillary tube until the rising force $(2\pi r \cos\theta \cdot \gamma)$ is balanced by the gravity of the liquid $(\pi r^2 \rho g h)$, as shown in Figure 3.6,

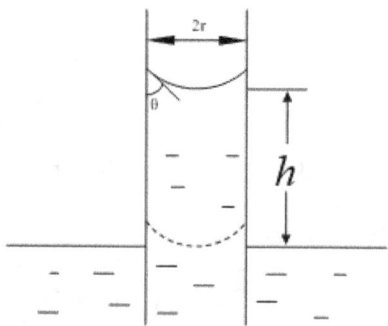

Figure 3.6. Surface tension measured by capillary rising method.

i.e., $2\pi r \cos\theta \cdot \gamma = \pi r^2 \rho gh$, or

$$\gamma = \frac{\rho ghr}{2\cos\theta} \tag{3.21}$$

where h is the liquid rising height. γ is the surface tension, ρ is the density of the liquid, r is the radius of the capillary, g is the acceleration due to gravity, and θ is the contact angle. Since ρ, g, and r are known, surface tension can be calculated by measuring the height to which the liquid rises.

For complete wetting, $\theta = 0°$, we have $\gamma = \frac{\rho ghr}{2}$.

3.5.2 *Du Noüy ring method*

The Du Noüy ring method is a traditional technique used to measure the surface tension of a liquid. In this method, a platinum wire ring is suspended from a torque scale. As the torque wire of the scale is gradually turned, the ring slowly lifts from the surface of the liquid, creating a circular film of liquid, as illustrated in Figure 3.7. The force required to raise the ring from the liquid's surface is measured and is directly related to the surface tension of the liquid. A measurement system for the Du Noüy ring method is depicted in Figure 3.7.

The Du Noüy ring method consists of eight steps, as illustrated in Figure 3.8.

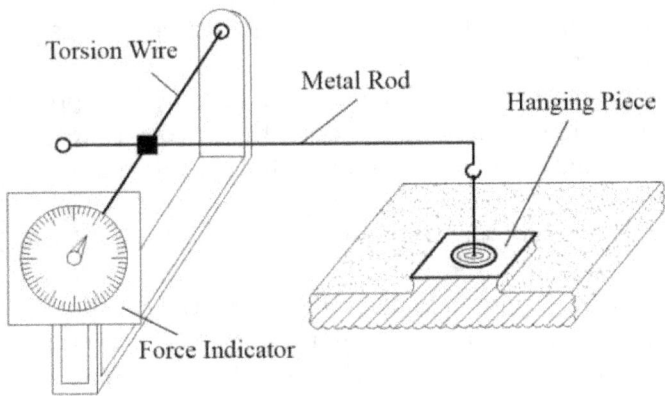

Figure 3.7. Surface tension measured using the Du Noüy ring method.

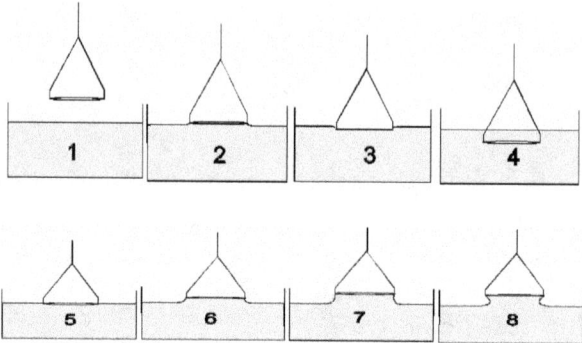

Figure 3.8. The Du Noüy method and the process of measurement.

(1) When the ring is positioned above the liquid surface, the measurement instrument registers a reading of zero.

(2) When the ring approaches the liquid surface, a slight attractive force exists between the ring and the surface due to adhesive attraction.

(3) The ring experiences an upward reverse force due to the slight action of surface tension. Consequently, the pulling force measured by the instrument is negative.

(4) When the ring immerses in the liquid surface, surface tension will dissipate. The weight of the ring overcomes buoyancy, resulting in a downward force transmitted through the wire. Consequently, a slight positive pull is detected by the instrument.

(5) When the ring is pulled, it will start to measure the pulling force.

(6) The force measured by the instrument increases as the ring is pulled upward.

(7) The state indicates that the maximum force is approaching.

(8) After reaching the maximum force, the force will experience minimal attenuation until the liquid film is ruptured.

The data curve of the pull is shown in Figure 3.9 corresponding to the eight steps above.

When the ring suddenly detaches from the surface film (while maintaining the metal lever in a horizontal position), the maximum pulling force F is equal to both the weight of the liquid being lifted, mg, and the surface tension acting around the ring. Since the liquid film has two sides, both inside and outside, the total circumference

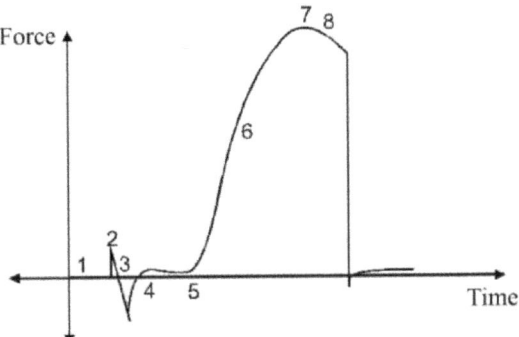

Figure 3.9. Curve of the measurement data.

of the ring is $4\pi R$:

$$F = mg = 4\pi R \cdot \gamma \qquad (3.22)$$

where m is the mass of the liquid which was pulled. R is the mean radius of the ring. The inner radius of ring is R'. The radius of platinum wire is r. The mean radius of ring is $R = R' + r$, which is shown in Figure 3.7. From equation (3.22), it can be obtained that

$$\gamma = \frac{F}{4\pi R} \qquad (3.23)$$

Therefore, if F is measured, γ can be obtained. However, since the pulled-up liquid is not cylindrical, equation (3.23) needs to be multiplied by a factor f:

$$\gamma = \frac{F}{4\pi R} \times f \qquad (3.24)$$

Many experiments indicate that the factor f is a function of $\frac{R^3}{V}$ and $\frac{R}{r}$, where V is the volume of the liquid pulled by the ring. V can be derived from the equation $F = mg = V\rho g$. The value of f can be found from Figure 3.10. The accurate value can be verified from related monographs.

3.5.3 *Maximum bubble pressure method*

One of the effective methods for determining dynamic surface tension is the "maximum bubble pressure method", or simply the "bubble

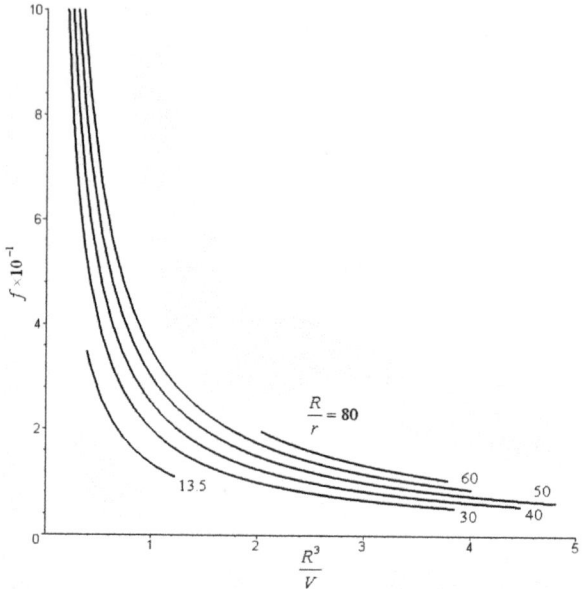

Figure 3.10. Curve of factor f.

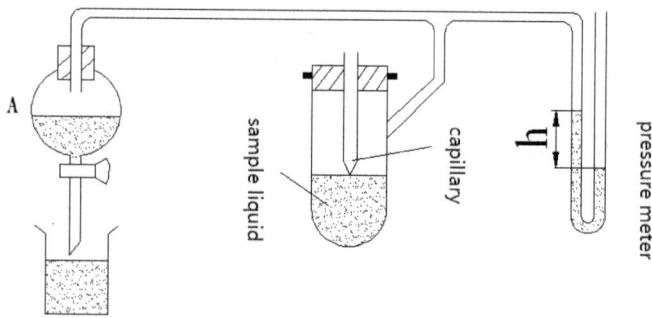

Figure 3.11. Equipment drawing of the maximum bubble pressure method.

pressure method". The equipment used for this method are illustrated in Figure 3.11. A bubble pressure tensiometer generates gas bubbles (e.g., air) at a constant rate and introduces them through a capillary tube that is submerged in the sample liquid, with the radius of the capillary already known. As the pressure inside the gas bubble continues to increase, the maximum pressure is achieved when the bubble takes on a completely hemispherical shape, and its radius,

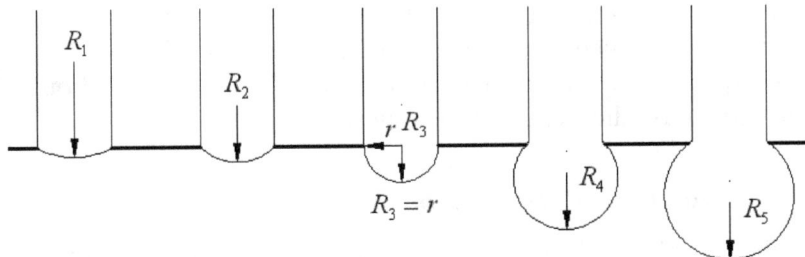

Figure 3.12. The change in the curvature radius of a bubble.

R, corresponds precisely to the radius of the capillary, r, as depicted in Figure 3.12. The height of liquid in the pressure meter is h when the pressure difference between the inside and outside of the bubble reaches its maximum, Δp_{max}, which can be written as

$$\Delta p_{\text{max}} = \rho g h \qquad (3.25)$$

The Δp_{max} is proportional to surface tension of the liquid and inversely proportional to the curvature radius, as given by

$$\Delta p_{\text{max}} \propto \frac{\gamma}{r} \qquad (3.26)$$

or

$$\Delta p_{\text{max}} = \frac{K\gamma}{r} \qquad (3.27)$$

The constant quantity K is equal to 2, which is determined using the Young–Laplace equation:

$$\Delta p_{\text{max}} = \frac{2\gamma}{r} = \rho g h \qquad (3.28)$$

or

$$\gamma = \frac{r}{2} \rho g h \qquad (3.29)$$

When two kinds of liquids, whose surface tensions are γ_1 and γ_2, are measured by the same capillary and pressure meter and the corresponding heights of the liquids are h_1 and h_2, it can be obtain from equation (3.29) that

$$\frac{\gamma_1}{\gamma_2} = \frac{h_1}{h_2} \qquad (3.30)$$

Therefore, when K is unknown, the surface tension of the sample liquid can be obtained from the liquid whose surface tension is known.

This method is widely utilized due to its advantageous characteristics. For instance, the measurement process is brief and employs simple equipment, and it does not require information about the contact angle or the density of the liquid.

3.5.4   *Wilhelmy plate method*

A Wilhelmy plate is a thin device used to measure the equilibrium surface or interfacial tension at the interface of a gas–liquid or liquid–liquid phase. The Wilhelmy plate typically measures a few square centimeters and is often constructed from glass or platinum, which is roughened to ensure complete wetting. In this method, the plate is oriented perpendicular to the interface, and the force exerted on it is measured, as illustrated in Figure 3.13.

When a plate is immersed in a liquid, the liquid will either ascend (in the case of a lyophilic liquid) or descend (in the case of a lyophobic liquid) along the vertical wall of the plate. The Wilhelmy plate measures the force F_W exerted by the liquid on the plate, which can be either a pull or push force, to determine the contact angle or surface tension. When the system reaches equilibrium, the interface of the gas–liquid–solid phases will achieve a mechanical balance:

$$F_{\mathbf{w}} = \gamma_{\mathbf{LV}} C \cdot \cos\theta - F_{\mathbf{b}} \qquad (3.31)$$

where $\gamma_{\mathbf{LV}}$ is the surface tension of the liquid, C is the length of line where the plate comes into contact with the liquid surface, θ is the contact angle, and $F_{\mathbf{b}}$ is the floatage which exists when the liquid

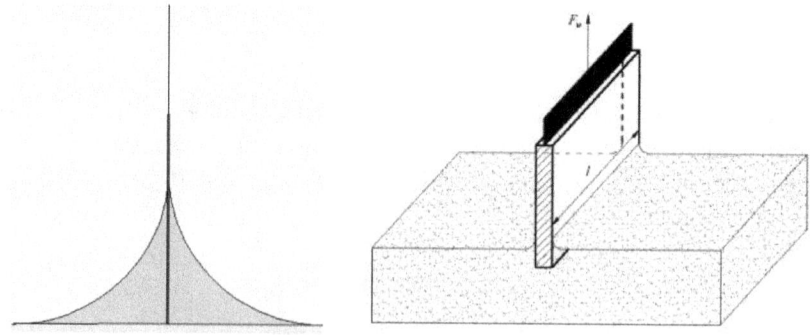

Figure 3.13.   Wilhelmy plate.

exerts on the plate. Since the two surfaces are wet, C is equals to $2l$. l is the width of the plate.

If the surface tension γ_{LV} is known, equation (3.31) can be changed into

$$\cos \theta = \frac{F_w + F_b}{\gamma_{LV} \cdot C} \qquad (3.32)$$

so as to derive the contact angle θ. If the contact angle θ is known, equation (3.31) can be changed into

$$\gamma_{LV} = \frac{F_w + F_b}{\cos \theta \cdot C} \qquad (3.33)$$

so as to get the surface tension γ_{LV}.

3.5.5 *Stalagmometeric method*

The stalagmometric method involves using a stalagmometer to measure surface tension. The stalagmometer, also known as a stactometer or stalogometer, is a capillary glass tube with a widened middle section, as illustrated in Figure 3.14. The bottom of the device is narrowed to allow the fluid to exit the tube in the form of a drop. The design of the stalagmometer ensures that the volume of each drop is consistent. During experiments, drops of the specific fluid flow slowly from the tube in a vertical direction. The drops that hang from the bottom of the tube begin to fall when their volume reaches a maximum value, which is determined by the characteristics of the

Figure 3.14. Stalagmometer.

solution. At this moment, the weight of the drops is in equilibrium with the surface tension. Based on Tate's law,

$$mg = 2\pi r \gamma \tag{3.34}$$

The drop falls when the weight (mg) is equal to the circumference ($2\pi r$) multiplied by the surface tension (γ). The surface tension can be calculated when the radius of the tube (r) and the mass of the fluid droplet (m) are known. Additionally, when the surface tension is proportional to the weight of the droplet, a reference fluid — typically water — can be used for comparison with the target fluid:

$$\frac{m_1}{\gamma_1} = \frac{m_2}{\gamma_2} \tag{3.35}$$

In the equation, m_2 and γ_2 are the mass and the surface tension of the reference fluid, while m_1 and γ_1 are the mass and the surface tension of the target fluid. If we take water as a reference, then

$$\gamma_1 = \gamma_{\mathbf{H_2O}} \cdot \frac{m}{m_{\mathbf{H_2O}}} \tag{3.36}$$

If the surface tension of water is known, we can calculate the surface tension of the target fluid using the equation. The more accurately the weights of the drops are measured, the more precisely we can determine the surface tension from equation (3.36).

The equipment used in this method to measure the weight of the liquid includes a stalagmometer, a beaker, a spherical tube, a weighing bottle with a lid, a weighing scale with a precision of one-thousandth, distilled water, a fluid with unknown surface tension, and a thermometer. The steps of the experiment are as follows:

(1) Place a clean and dry stalagmometer vertically on the experimental plate.
(2) Weigh the mass of the weighing bottle.
(3) Obtain some distilled water using a beaker. Cover the top of the stalagmometer with a rubber tube. Introduce water into the stalagmometer until the water's surface reaches the wider middle section of the stalagmometer.
(4) Remove the rubber ball to allow the droplet to fall freely, and collect 20 drops in the weighing bottle.
(5) Measure the weight of the weighing bottle containing 20 drops.

(6) Please empty the weighing bottle and dry it for the next measurement.
(7) Change to the target liquid and repeat Steps 2 through 6.
(8) Measure the temperature of the laboratory using a thermometer, and examine the surface tension of water at the specified temperature. Then, substitute this value into equation (3.36) to calculate the surface tension of the target liquid at that temperature.

The equipment used in this method for counting droplets include a stalagmometer, a beaker, a spherical tube, a weighing bottle with a lid, a precision balance with a thousandth of a gram accuracy, distilled water, a fluid with unknown surface tension, and a thermometer.

There are two indicator lines on the stalagmometer. The upper line is located above the wide middle section, while the lower line is positioned below it. The volume between these two lines is V. The density of liquid is ρ. The mass is

$$m = V \cdot \rho \qquad (3.37)$$

Set the two kinds of dropping liquids in the same volume. The number of drops n is accounted for by the volume V. Then, the mass of each drop is

$$\bar{m} = \frac{m}{n} = \frac{V \cdot \rho}{n} \qquad (3.38)$$

The mass in equation (3.36) is replaced with $\bar{m}$. Then, the surface tension of the target liquid is

$$\gamma_1 = \gamma_{H_2O} \cdot \frac{\rho}{\rho_{H_2O}} \cdot \frac{n_{H_2O}}{n} \qquad (3.39)$$

The steps of this experiment are as follows:

(1) Pour distilled water into the stalagmometer until the water level reaches the upper line. Open the stalagmometer to allow the water to drop into the weighing bottle until the water level reaches the lower line. Close the stalagmometer and record the number of drops of water.
(2) Empty the weighing bottle and dry it thoroughly for the next measurement.

(3) Change to the target liquid and repeat Steps 1 and 2.
(4) Measure the temperature of the laboratory using a thermometer and assess the surface tension of water at the specified temperature. Substitute this value, along with the densities of water, into equation (3.39) to calculate the surface tension of the target liquid.

3.6   Methods of Measuring the Surface Tension of Solids

The surface tension of a liquid is characterized by an increase in the surface area. In contrast, the surface tension of a solid is influenced by its surface properties. However, solids differ significantly from liquids; the atoms and molecules in a solid cannot move freely as they do in a liquid. Consequently, there is currently no direct and reliable method for measuring the surface tension of solids. Instead, an indirect method is typically employed to assess the surface tension of solids, or it can be estimated using theoretical approaches.

In the early years, Soviet scientists used a method in which a crystal is struck with a blade. The work done by splitting the crystal can be determined when the crystal splits and a new surface is formed. For example, in this method, the measurement value of NaCl is $150 \, \mathrm{mN \cdot m^{-1}}$. This result is consistent with the calculation of crystal lattice. Two methods which are commonly used are introduced in the following sections.

3.6.1   *Critical surface tension method*

Currently, the critical surface tension method is a common method which is widely used to measure the surface tension of solids.

In the early 1960s, Zisman and other scientists discovered that when a series of liquids is applied to a macromolecule, the surface tension of the liquid is proportional to the cosine of the contact angle θ, as illustrated in Figure 3.15. The surface tensions of the series of liquids are known, and the surface tension of the solid surface is relatively low.

If the line in Figure 3.15 is extended to $\cos \theta = 1 (\theta = 0°)$, the corresponding surface tension of the liquid is the critical surface tension γ_c. For example, the surface tension of polythene (γ_c) is $31 \mathrm{mN \cdot m^{-1}}$.

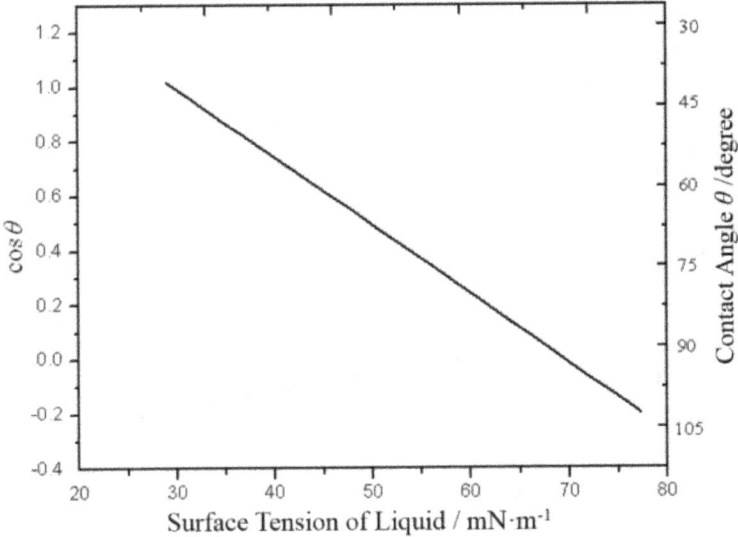

Figure 3.15. Zisman curve of polythene surface.

Table 3.4. Critical surface tension of some kinds of solids.

Solid	Perfluorinated dodecanoic acid	PTFE	Napht- halene	Hexa- decane	Poly- ethylene	Poly- styrene	PVC	Nylon
γ_c	6	18	25	29	31	33–43	39	42–46

The physical meaning of γ_c is that if the surface tension of it (γ) is smaller than γ_c, the liquid can wet the surface of polythene. On the contrary, it cannot wet the solid surface. The critical surface tensions of some kinds of solids are listed in Table 3.4. The γ_c increases according to the increase of the surface polar.

3.6.2 Calculating the surface tension of solid utilizing the surface tension of high polymer liquid or melt

In general, the surface tension of a liquid decreases as the temperature increases. The surface tension of high polymer melts is first measured at various temperatures and then extrapolated

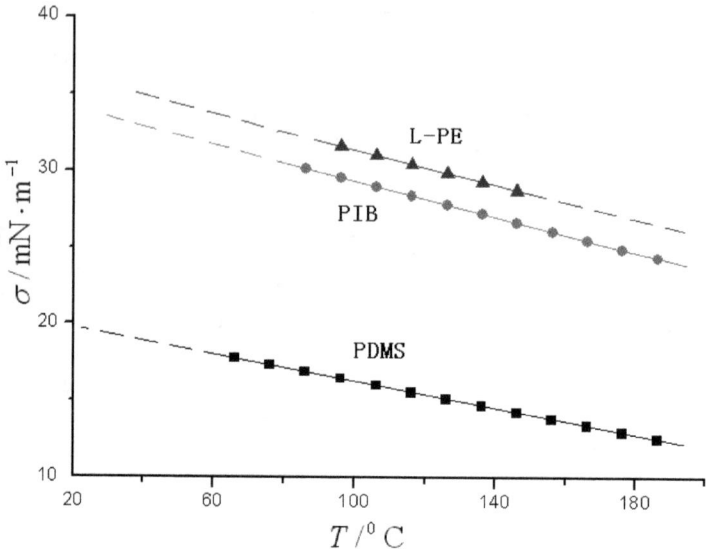

Figure 3.16. $\gamma-T$ curve of some kinds of high polymers.

to determine the surface tension of the corresponding solid state at a specific temperature. For example, the relationships between the surface tension and temperature of L-PE, PDMS, and PIB are illustrated in Figure 3.16.

According to the figure above, if the $\gamma-T$ curve is extended to 20°C, γ of L-PE is $35.7\,\mathrm{mN\cdot m^{-1}}$, that of PIB is $34.0\,\mathrm{mN\cdot m^{-1}}$, and that of PDMS is $19.8\,\mathrm{mN\cdot m^{-1}}$. Obviously, when using this method, the effects of phase change on the surface tension must be noted. Although this influence is always considered very small, the result of this method is approximate.

3.6.3   *Estimation method*

A solid is composed of atoms or molecules that occupy fixed positions. In theory, if the forces within the crystal lattice of the solid are understood, the surface tension of the solid can be calculated. For instance, if the crystal structure of a metal and the coordination number of the surface atoms are known, one can determine the superficial energy of each atom as well as the superficial energy per unit area (i.e., the surface tension). The coordination number of surface atoms refers to the number of atoms surrounding a single atom in the superficial layer, excluding the influence of atoms in the second layer. In this method, the surface tension of Cu is $1.5 \text{N} \cdot \text{m}^{-1}$, which is consistent with the test value. However, this method is not suitable for ionic crystals.

3.7 Example of Surface Tension Measurement with Interfacial Tension Instrument ZYW-200B

ZYW-200B is an automatic-control instrument designed for measuring interfacial tension and surface tension, utilizing the Du Noüy ring method. A key feature of this instrument is its reliance on a physical measurement approach rather than traditional chemical methods, making it user-friendly. With ZYW-200B, users can quickly and accurately measure the surface tension or interfacial tension of various liquids. Additionally, the measurement results are automatically displayed and recorded, and the measurement curve can be easily stored or printed.

3.7.1 *Structure and working principle*

This instrument consists of a torque wire, a differential transformer, a platinum ring, and a lifting mechanism, as illustrated in Figure 3.17. Its operating principle is as follows.

When measuring surface tension, the platinum ring is first immersed in the target liquid. Then, the filled glassware on the salver

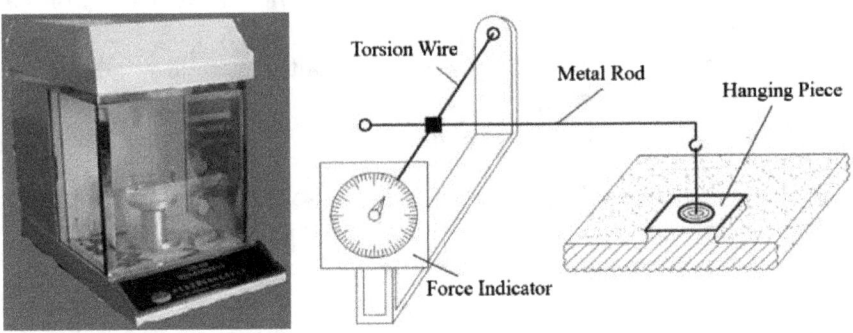

Figure 3.17. Automatic tension device and its schematic diagram.

is lowered by the transmission system. As the glassware descends, the film between the platinum ring and the liquid is stretched. Consequently, the platinum ring experiences a downward force, causing the torque wire to rotate through a lever mechanism. As a result, the magnetic core of the differential transformer ascends while the platinum ring descends. The coil of the differential transformer induces a specific voltage. Therefore, the amount of film deformation is converted into voltage, which is then translated into the corresponding tension through computer calculations and displayed automatically. As the film is gradually stretched, the tension increases until the film ruptures. During this procedure, the maximum value is the measurement value of the tension γ'. While this value is multiplied by the liquid correction factor F. Then, the actual tension value γ will be obtained by $\gamma = \gamma' \cdot F$. The factor is determined by γ', the density of the liquid, the radius of the platinum wire, and the radius of the platinum ring.

3.7.2 *Preparing for the experiment*

1. Put the instrument on a stable table board carefully. Keep it horizontal by adjusting the knob on the base. Keep the room temperature at $20 \pm 5°C$.
2. Open the top lip and check out the following things:

   - The magnetic core of the differential transformer should be in the normal position, where it cannot come into contact with the wall but hangs freely.
   - The torque wire should be tensioned. If not, adjust the bolts at the two ends.
   - The level arm should be kept horizontally. If not, adjust the two ends of the torque wire.
   - After checking, close the upper lip.

3. Insert the power plug into an external socket reliably grounded. Turn on the power button and stabilize it for 30 minutes. Clean the platinum ring and the glass. First, the platinum ring is washed with petroleum and then with acetone. Next, the ring is heated and dried in the oxidation flame of a gas lamp or an alcohol lamp. When dealing with the platinum ring, avoid deforming the ring.

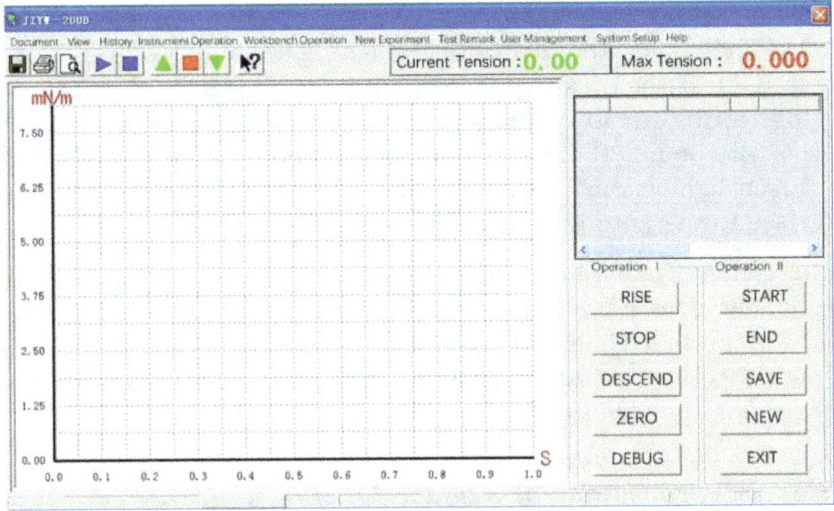

Figure 3.18. The software interface.

Then, hang the ring on a small hook of the lever arm. When it is stabilized, start the experiment.

4. Run the program and enter the experimental preparation interface, as shown in Figure 3.18. After inputting the parameters and confirming, the main interface will appear. At this point, the machine automatically becomes zero, and the adjustment range is $\pm 10.0\,\mathrm{mN/m}$. If it is in this range, the experiment can be started. If not, the screen will automatically display a message. Open the cover and adjust the lift rod until it reaches the adjustment range (preferably transferred into a value of zero or very small). Then, press the zero key to set zero.

3.7.3 Calibrating the instrument

The mass method is used to calibrate the instrument. The procedure of calibration is described as follows.

A small piece of paper is placed on the platinum ring after the instrument is installed. Click on the instrument calibration option in the main interface menu. A measurement value is displayed. If the value is not zero after stabilization, click the zero button. Then, put a weight of 1000 mg on the paper. At this point, the value displayed should be 81.7 mN/m (the relative error of the indicated value is ±1%). Then, click the menu for the actual value of the weight and select 1000. If the green light is on after stabilization, click the exit button and return to the main interface to measure the surface tension of the liquid. If the red light is on, click the calibration until the green light is on. Click the exit button and return to the main interface to measure the surface tension of the liquid.

Note: Generally, if the tension is in the allowable error range, the instrument does not need to be adjusted. When the instrument is used for the first time or is stored for a long period of time before using, the test error is considered to be bigger. Therefore, the instrument should be calibrated.

3.7.4 Measurement of surface tension and interfacial tension of liquid

3.7.4.1 *Measurement of surface tension of liquid*

(1) Pour the liquid sample, which is adjusted to 25°C, into a beaker, and keep the height of the liquid surface at 20–25 mm. Then, put the beaker at the center of the tray, as shown in Figure 3.19.

Figure 3.19. The flying ring and the beaker.

(2) Adjust the zero. The software interface is shown in Figure 3.18. The adjustment is done by clicking the "zero" button. This step is only done before the first test when conducting many tests.

(3) By clicking the "up" button, the platinum ring will come into contact with the liquid. When the ring becomes immersed in the liquid to about 5–7 mm, the "stop" button is clicked.

(4) Click the "experiment start" button. The experiment begins and the tray and the target liquid begin to descend. The real curve is drawn and the maximum surface tension is displayed by the software, as shown in Figure 3.20.

(5) When the experiment is finished, the software will end the experiment automatically. It can also be ended by clicking the "experiment end" button.

(6) Repeat Step 3 for the next experiment. The software of this instrument allows one to conduct eight experiments for one sample continuously and process the data.

(7) Click the "save" button to save the experiment data. Click the "browse data" button to check or print the data in the experiment data window.

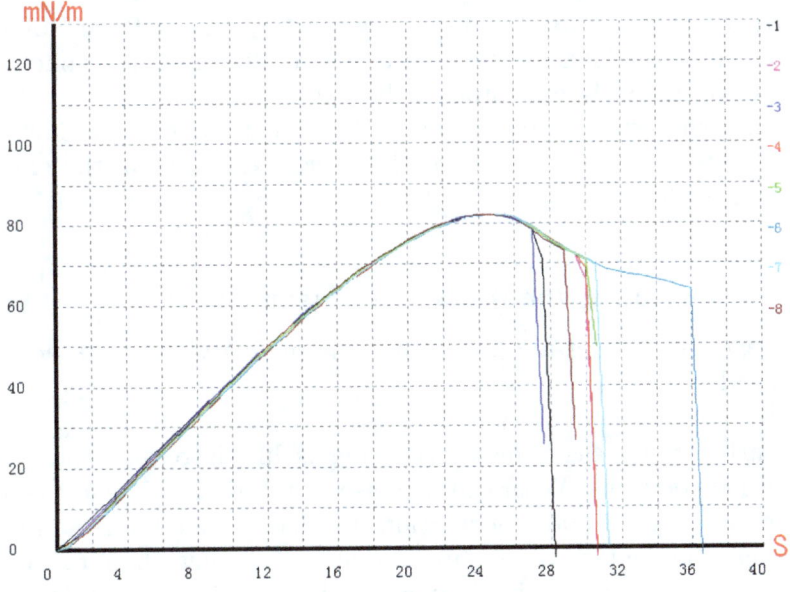

Figure 3.20. Real curve drawn by ZYW-200B.

3.7.4.2   *Measurement of the interface tension of liquid (taking the interfacial tension of petroleum products to water as an example)*

(1) Pour quantificational distilled water at 25°C in a beaker. Keep the height of the liquid surface at 20–25 mm. Then, place the beaker at the center of the tray.

(2) Lift the tray so that the platinum ring can immerse into the liquid with a height of 5–7 mm.

(3) Pour the sample liquid, which has been adjusted to 25°C, into the distilled water. The height of the sample is about 10 mm. Note that the sample liquid cannot come into contact with the ring, and the ring cannot come into contact with the interface of oil–water. Keep the surface of oil–water for 30 s.

(4) Click the "experiment start" button. The experiment begins and the tray and the target liquid begin to descend. The real curve is drawn and the maximum surface tension is displayed by the software.

(5) When the experiment is finished, the software will end the experiment automatically. It can also be ended by clicking the "experiment end" button.

(6) Repeat Step 3 for the next experiment. The software of this instrument allows one to conduct eight experiments for one sample continuously and process the data.

(7) Click the "save" button to save the experiment data. Click the "browse data" button to check or print the data in the experiment data window.

3.7.5   *Correcting the measured tension*

In the Du Noüy ring method of measurement, two aspects need to be considered:

(1) During the measurement, the ring is lifted so that the liquid surface deforms. With the increase in the rising displacement of the ring, the liquid deformation increases. Therefore, the radius from the center to the break point is smaller than the mean radius of the ring. This influence is indicated by the ratio of the radius of the ring and the radius of the platinum.

(2) There is a little amount of liquid adhering to the bottom of the ring. This influence can be expressed by a function.

According to the two situations above, the measured tension γ' should be multiplied by a correction factor F, which can be derived from the actual tension $\gamma \cdot \gamma = \gamma' \cdot F$. According to ASTM D971, the correction factor of the device can be calculated using the equation

$$F = 0.7250 + \sqrt{\frac{0.01452 \times \gamma' \div C^2 \div (D-d) + 0.04534}{-1.679 \times r \div R}} \qquad (3.40)$$

where γ' is the displayed value (mN/m), C is the circumference of the ring, R is the radius of the ring, D is the density of the bottom phase at 25°C (g/ml), d is the density of the top phase at 25°C (g/ml), and r is the radius of the platinum wire.

If the two measured phases are gas and liquid, D is the density of liquid and d is the density of gas.

3.7.6 *Note*

If you want to check the historical curves, click the "open" button under the "file menu" or click the "open curve" button. If you wish to test a different sample liquid, click the "experiment prepare" button to return to the parameter interface and input new parameters for the new experiment.

During the experiment, the instrument should be positioned on a vibration-free platform, preferably an isolation platform. It is essential to ensure that the instrument is leveled before commencing the experiment. Once the experiment is complete, turn off the power, remove the platinum ring, and clean the instrument. It is important to note that interference from high-power electrical devices and wireless communication equipment is prohibited during the experiment. Additionally, the laboratory housing the instrument should be equipped with temperature control systems to maintain a constant temperature. Measures should also be taken to prevent contamination from harmful gases.

If the repeatability of the experiment is poor, there may be several underlying reasons:

(1) The platinum ring and glass cup may not have been cleaned.
(2) The liquid may have expired.
(3) The magnetic core of the differential transformer may be in contact with the wall.
(4) The torque wire may be loose.

3.8   Elasticity of Liquids

There are several methods to measure the surface tension of liquids, including the capillary tube method, maximum bubble method, pull-out method, pendant-drop method, and ring method. Among these, the ring method is the most commonly used. There are specific standards that must be followed when employing this method. The ring method involves immersing a ring in the liquid and then applying a force to pull the ring upward until it separates from the liquid. The surface tension is calculated as the maximum force divided by twice the perimeter of the ring. From the measurement process, it can be observed that surface tension represents the tensile limit or strength of the liquid. When measuring the surface tension of a liquid using the ring method, the force required to lift the ring increases with the height of the liquid column being raised by the ring. Once the lifting force reaches its maximum value, any further increase in the height of the liquid column will cause the force to decrease rapidly, ultimately becoming zero when the liquid column is broken. This measurement process reveals that the force lifting the ring is directly proportional to the displacement of the liquid being pulled. To study the characteristics of a liquid, a lifting plate (silicon plate) can be employed to conduct a pulling test with purified water. When the lifting plate is completely wetted by the liquid, lifting the plate with an upward force will generate a counteracting force from the liquid. This upward force arises from the adhesion between the liquid and the solid plate, as well as the cohesive forces within the liquid itself. The solid plate must overcome this force to detach from the liquid surface. The magnitude of this force can be quantified using a measuring device. When lifting the plate upward, a specific height of liquid is drawn up, forming an inward-curving liquid column. When the height of the lifted liquid column is minimal — prior to the lifting force reaching its maximum value — the liquid exhibits properties akin to solid elasticity.

This liquid column undergoes repeated deformation through loading and unloading, a phenomenon referred to as liquid elasticity. This behavior is also observed during the measurement of surface tension using the ring method.

3.8.1 *Linearity in liquid tensile process*

First, when doing the liquid tensile test, use the lifting plate to pull the liquid. Test the relationship between the force for lifting the liquid column and the height of the pulled-up liquid column. Before the lifting force reaches its maximum value, the height of the lifted liquid column is very small. If using d to represent the height of the pulled-up liquid and f to represent the corresponding lifting force, the lifting force for the first height d_{s1} is f_{s1}. In the same way, the lifting force for the second height d_{s2} is f_{s2}, the lifting force for the nth height d_{sn} is f_{sn}. By marking the points on the coordinates, the previous points (d_{s1}, f_{s1}), $(d_{s2}, f_{s2}) \cdots (d_{sn}, f_{sn})$ appear in a perfect linear relationship and satisfy the linearity equation $f = a + kd$. Upon unloading the plate, the liquid column declines to the mth height, the corresponding force is f_{jm}. By marking the points on the coordinates, the points (d_{j1}, f_{j1}), $(d_{j2}, f_{j2}) \cdots (d_{jm}, f_{jm})$ also appear in a perfect linear relationship and satisfy the previous linearity equation law, and every parameter is the same as the above, which means the tensile force of the liquid and the height of the lifting liquid column appear to have a linear relationship. The curve between the tensile force and displacement is shown in Figure 3.21. The silicon plate is in the size of 17.725 mm · 17.725 mm and the liquid is purified water. The test has been done six times and shows that before reaching the maximum force, the force and displacement appear in a linear relationship, and the repeatability of each-time curve is good.

3.8.2 *Recoverability in liquid tensile process*

The height of the lifted liquid column is very small (before the lifting force reaches its maximum value), and the liquid above shows not only the linearity property but also a good recoverability if the liquid is pulled up by the plate repeatedly. Suppose in the process of pulling up the liquid column, f_0 is the force of lifting the plate and d_0 is the

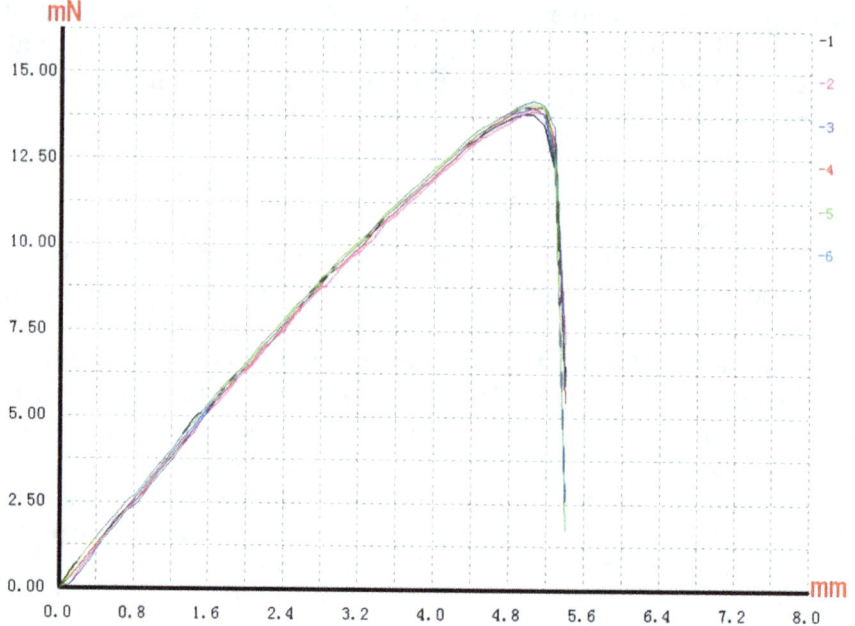

Figure 3.21. Force–displacement curve of silicon plate pulling up liquid.

corresponding height of the pulled-up liquid column. If the plate is lifted continuously, the force will continue to increase. If the plate goes downward to the previous height d_0 by unloading (before the tensile force reaches its maximum value), the liquid bridge force, which is equal to the force on the lifting plate, will also return to f_0. The same occurs with the pulling force when loading to the same height. Therefore, this force is recoverable.

Further tests indicate that the liquid pulling force remains constant when both the liquid column height and the plate area are fixed, regardless of whether the system is in a loading or unloading state. When the liquid column is subjected to repeated loading and unloading, this behavior demonstrates that the liquid possesses good elasticity. Additionally, when measuring the surface tension of the liquid using the ring method, a similar phenomenon is observed. Figure 3.22 shows the curve between the tensile force and displacement by using a silicon plate with a size of 17.725 mm · 17.725 mm to pull up purified water in the beaker repeatedly (loading and unloading). To distinguish between the loading and unloading processes, a

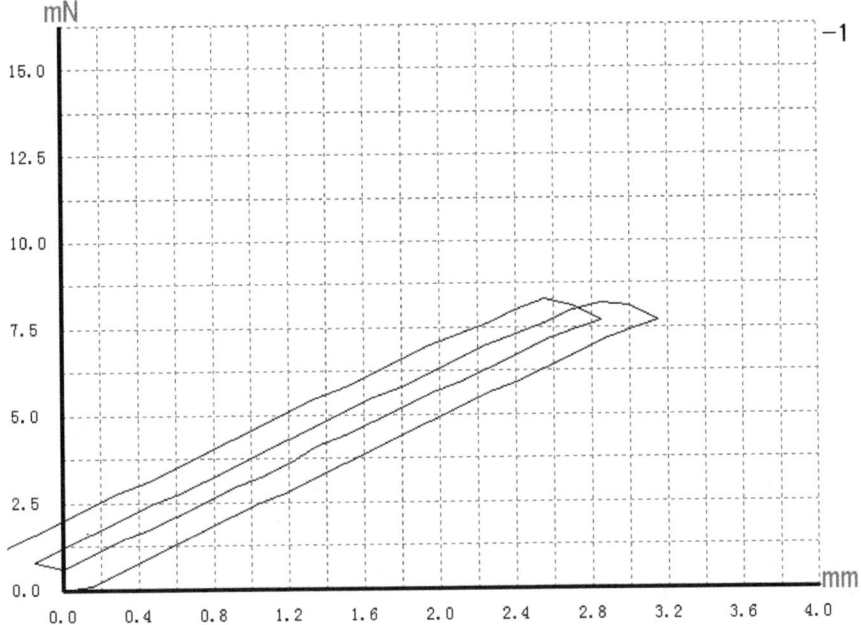

Figure 3.22. Force–displacement measuring curve of pulling and recovering the liquid repeatedly.

transmission gap is maintained during the conversion between these two phases. This transmission gap results in an approximately horizontal segment at the bend of the force curve, as illustrated in the figure A control software can be utilized to eliminate the influence of the test device gap and to generate the lifting and declining force–displacement curve, as shown in Figure 3.23. In each stage (referring to the distinct lifting and declining phases), not only is the force–displacement relationship linear, but the fitted linear equations for the different stages also share the same parameters through curve fitting. Suppose the fitted equation for each stage of linearity is represented as $Y = A + B \times X$, based on tests conducted with silicon and purified water in a beaker. The OriginPro 7.5 software was utilized to analyze the corresponding data for each stage, resulting in the parameters of the linear equation, as presented in Table 3.5. In the table, the parameters A and B are remarkably consistent, and the dispersion error is minimal, indicating a strong linearity property. The similarity of the parameters suggests a significant regularity in

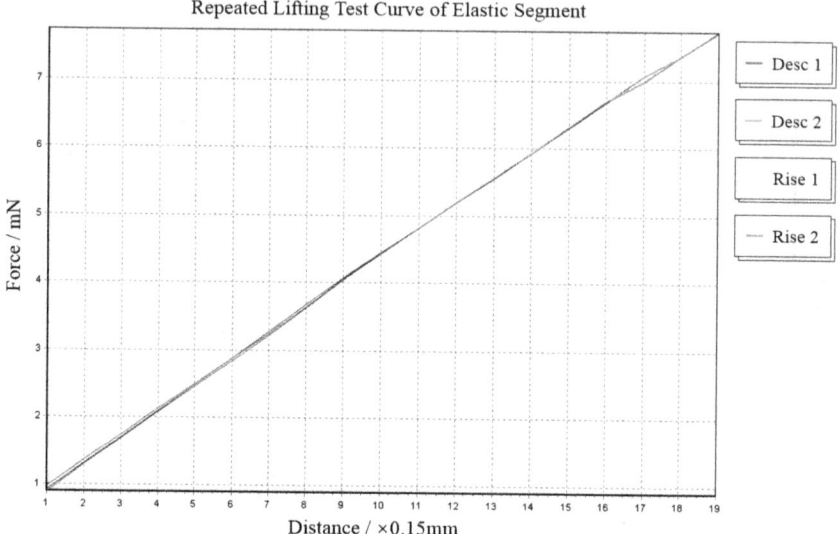

Figure 3.23. Force–displacement measuring curve of pulling and recovering the liquid repeatedly with device effect removed.

Table 3.5. Data list of each-stage linear fit equation of silicon plate pulling purified water repeatedly.

	A		*B*			
	Value	**Error**	**Value**	**Error**	**R**	**S_D**
Lifting 1	0.56333	0.0193	2.54339	0.01128	0.99983	0.04041
Lifting 2	0.61912	0.01568	2.52129	0.00917	0.99989	0.03283
Declining 1	0.61053	0.02142	2.52807	0.01253	0.99979	0.04486
Declining 2	0.59351	0.02115	2.5317	0.01237	0.9998	0.04429
Average	0.5966	0.0194	2.5311	0.01133	0.99983	0.040598
Standard deviation	0.0246	0.0026	0.0093	0.00134	0.00004	0.004800

the data. Furthermore, the consistency observed in both the lifting and declining processes demonstrates a high degree of recoverability. As illustrated in the data curve in Figure 3.23, the tensile force and displacement exhibit a strong linear relationship and excellent recoverability during the processes of extension and recovery.

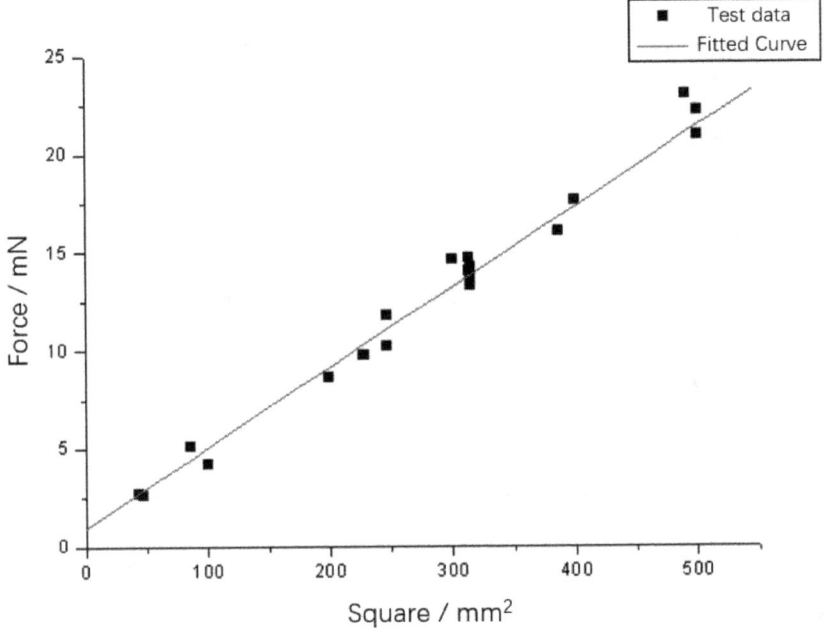

Figure 3.24. The maximum tensile force of liquid per unit area has nothing to do with the area.

3.8.3 The maximum tensile force of liquid on per unit area is independent of the area

Tests indicate that when the areas of the plates vary, the maximum tensile force exhibits different values. Let the maximum tensile force be denoted as f_{max} and the corresponding maximum height of the liquid column be d_{max}. Figure 3.24 illustrates the relationship between the maximum tensile force of the liquid and the area of the silicon plate. The curve demonstrates a strong linear correlation between the maximum force and the area (specifically, the area of the liquid column), with minimal dispersion error.

The maximum force per unit area can be figured out by dividing the maximum force of the liquid column area, as shown in Figure 3.25. In Figure 3.25, the curve is a horizontal line, which shows that the maximum tensile force of liquid per unit area has nothing to do with the area, which means it is independent of the area.

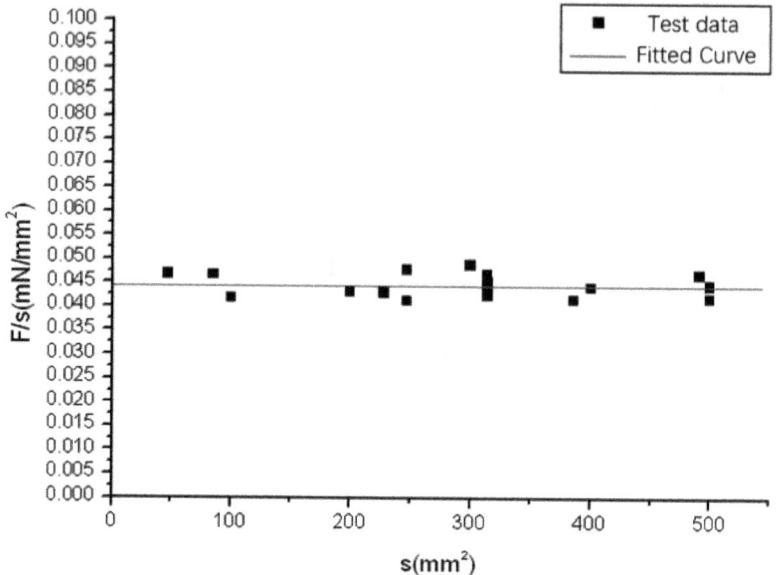

Figure 3.25.   The maximum tensile force of liquid per unit area has nothing to do with the area.

3.8.4   *The Hook's law of liquid tension*

In order to describe elasticity of liquids, the tensile force per unit area is defined as liquid tension, which is represented by σ_l. That is $\sigma_l = f/A$. The ratio between the height of the pulled-up liquid column and the maximum height of the liquid column is defined as the liquid tensile ratio, which could also be called the relative tensile height. Use ε_l to represent the height and $\varepsilon_l = d/d_{\max}$.

From Section 3.8.1, the tensile force is in a linear relationship with the height of the pulled-up liquid column. Set the height of the pulled-up liquid column as zero when $f_0 = 0$, that is, pulling from the origin of the coordinates. The tensile force f and the height d of the pulled-up liquid column have a linear relationship as follows:

$$f = kd, \quad d \le d_{\max} \tag{3.41}$$

When the tensile force reaches the maximum value $f_{\max}$, the corresponding height of the liquid column is $d_{\max}$. According to the previous formula,

$$f_{\max} = kd_{\max} \tag{3.42}$$

It could be seen from Section 3.8.3 that the maximum force of liquid columns with different areas is in a linear relationship with the areas of the liquid columns. Suppose the slope of the linear equation is E_l and the area of the liquid column is A. Since the silicon does not touch the liquid when the liquid column area is zero ($A = 0$) and the tensile force is zero ($f = 0$), the linear equation between the maximum force and the area can be written as follows:

$$f_{max} = E_l A \qquad (3.43)$$

According to equations (3.42) and (3.43), we have

$$k = \frac{A}{d_{max}} E_l \qquad (3.44)$$

Substituting equation (3.44) into (3.41) leads to

$$f = \frac{d}{d_{max}} E_l A \qquad (3.45)$$

Transform (3.45) into

$$\frac{f}{A} = \frac{d}{d_{max}} E_l \qquad (3.46)$$

Using the previous definition, equation (3.46) can be rewritten as

$$\sigma_l = E_l \varepsilon_l \qquad (3.47)$$

Equation (3.47) is an expression of the elasticity of liquids. Compared with Hook's law of solid materials, equation (3.47) could be called the Hook's law of liquid tension. In equation (3.47), E_l is equivalent to the elasticity modulus. Therefore, it is called the elasticity modulus of a liquid. Comparing with the elasticity of a solid, the tensile stress of the liquid σ_l corresponds to the tensile stress of the solid material. The tensile ratio of the liquid ε_l corresponds to the tensile strain of the solid material. The elasticity modulus of the liquid E_l corresponds to the elasticity modulus of the solid. It should be noted that tensile ratio is the ratio between the height of the pulled-up column and the maximum height of the pulled-up column, which is different from that of the solid. E_l only has a relation with the liquid property and has nothing to do with the area and material of the lifting plate.

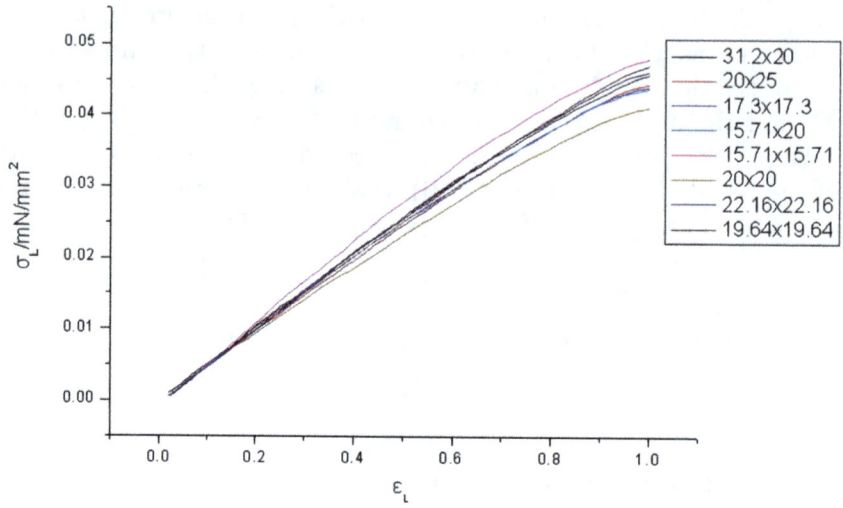

Figure 3.26. $\sigma_L - \varepsilon_L$ test curves of silicone plates with different areas.

Since equation (3.47) can be applied to both the loading and unloading processes, the liquid has a good linear elasticity property when the deformation is very small before reaching the maximum surface tension. Through a series of tests on the tensile force of the lifting plates with different areas, the test curves of different areas are obtained, as shown in Figure 3.26. $\sigma_L - \varepsilon_L$ of the liquid appears to exhibit very good linearity, and the stress–strain curves of silicone plates with different areas are basically coincident.

Chapter 4

Contact Angle and Wetting

4.1 Basic Concepts

4.1.1 *Contact angle*

In Figure 4.1, consider a liquid droplet that falls onto a solid surface. The liquid droplet will form a spherical crown with a defined volume. However, the bottom area of the droplet varies depending on the type of liquid, the properties of the solid surface, and the interactions between the liquid and the solid. The stronger the adhesion force between the liquid and solid, the larger the contact area will be. Conversely, the weaker the adhesion force, the smaller the contact area will be. When the contact area is larger, the angle between the surface of the droplet and the solid surface at the three-phase boundary (liquid, solid, and gas) is smaller. Conversely, when the contact area is smaller, the angle is larger. Therefore, this angle reflects the degree of interaction between the liquid and the solid surface, which is referred to as the contact angle.

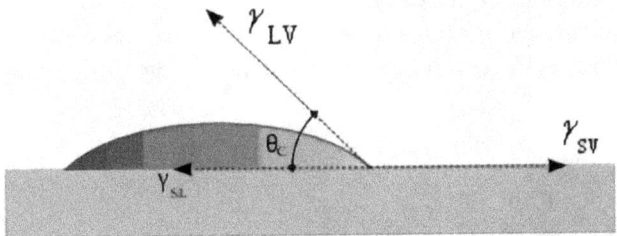

Figure 4.1. Contact angle.

The size of the contact angle depends on not only the liquid property and the solid surface properties but also the interaction between the two properties. According to their different properties, the contact angle can be any degree from 0° to 180°. Even for the same kind of liquid and solid, when there are differences in flatness, hardness, surface roughness, and chemical consistency of the solid surface, the contact angle will also be different. For analysis convenience, a specific idealized solid surface is often considered. The so-called ideal solid surface is considered to be flat, rigid, and smooth and to have uniform chemical properties but no contact angle hysteresis. If it does not possess the characteristics described above, it is the actual solid surface and cannot be called an ideal solid surface.

4.1.2   *Wetting*

Wetting phenomenon occurs when a liquid comes into contact and adheres to a solid surface. Wetting is the ability of a liquid to maintain contact with a solid surface, resulting from intermolecular interactions (adhesive force) when the liquid and solid are brought together. The degree of wetting (wettability) is determined by the force balance between the adhesive force and cohesive force. When a droplet of liquid falls on a solid surface, the adhesive force between the liquid and solid causes the liquid droplet to spread across the surface, while the cohesive force within the liquid causes the drop to ball up and avoid contact with the surface. Since the contact angle results from the combined action of the adhesive and cohesive forces, the contact angle also reflects the degree of wetting. Figure 4.2 illustrates the wetting behavior of various liquids on a solid surface. The figure indicates that a smaller contact angle corresponds to a greater propensity for droplet spreading and a stronger adhesion between the liquid and the solid surface.

A contact angle which is less than 90° (low contact angle) usually indicates that wetting of the surface is very favorable so that

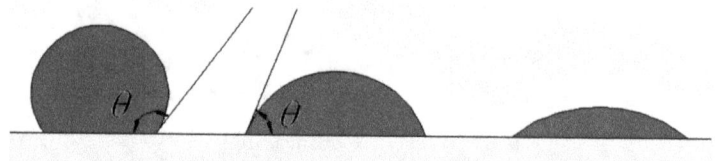

Figure 4.2.   Contact angle and wetting.

Table 4.1. The correlation between contact angle and the degree of wetting.

| | | Strength of | |
		Sol./Liq. interactions	Liq./Liq. interactions
Contact angle	**Degree of wetting**		
$\theta = 0^0$	Perfect wetting	Strong	Weak
$0^0 < \theta < 90^0$	High wettability	Strong	Strong
		Weak	Weak
$90^0 \leq \theta < 180^0$	Low wettability	Weak	Strong
$\theta = 180^0$	Perfectly non-wetting	Weak	Strong

the fluid will spread over a large area of the surface. Conversely, a contact angle greater than 90° (high contact angle) generally signifies that wetting of the surface is unfavorable, causing the fluid to minimize contact with the surface and form a compact liquid droplet. For water, a surface that is easily wetted is referred to as hydrophilic, while a surface that is not easily wetted is termed hydrophobic. Superhydrophobic surfaces exhibit contact angles greater than 150°, indicating almost no contact between the liquid droplet and the surface. Table 4.1 illustrates the relationship between contact angle and wetting.

Wetting is a common phenomenon. From the view of the phenomenon, it is also a process which can be thought of as a process of replacing one immiscible fluid with another on a solid surface. Therefore, the wetting phenomenon must also be associated with three phases. One of the phases must be a solid phase, and the other two phases are fluid phases. Normally, wetting indicates that a gas on the solid surface was replaced by a liquid. Therefore, solid properties should be concerned for wetting, especially the solid interfacial properties. The essence of the wetting phenomenon is the relationship between the solid–gas interface and the solid–liquid interface.

Since the wetting phenomenon involves solid–liquid interaction, this phenomenon involving a certain kind of liquid can also help us study some properties of the solid phase interface. For the interface phenomenon, the research on the solid surface or interface in reality is the most common one. Since this research is very difficult, compared with that on liquids, people rarely know any properties about the solid surface or interface. The wetting phenomenon is the macroscopic performance of microscopic molecular interactions between solid and liquid molecules at their interface. Therefore, a study on the wetting phenomenon will provide much indirect information and create conditions for people to understand the properties of solid surfaces.

Meanwhile, the wetting phenomenon can provide information about solid and liquid interactions. Under the precondition of known solid surface properties, a study on the wetting of unknown liquid adhered to the known solid surface can help us indirectly understand some surface properties of the unknown liquid. Since some kinds of liquids are soluble in water or interact with mercury, especially with mercury being harmful to the human body, we cannot obtain the basic surface force properties by using the direct interaction of those kinds of liquids with water or mercury. However, if there is no chemical reaction between such liquids and the known solid, by studying the wetting phenomenon of the liquids on a known solid surface, it is also possible to obtain some relevant basic properties about surface force.

In nature, there are two types of solid surfaces. Both of them can produce an effect with liquids. However, the effect strengths of these two types are different. Usually, they are called high-energy surface and low-energy surface. To show the relative interaction energy of a liquid with a solid surface, it must be jointly acted on with the solid bulk itself. Metals, glass, and ceramics are usually seen as hard solids because their ontologies are held together firmly through internal bonds (covalent bonds, ionic bonds, metallic bonds, etc.). To break them, a very high energy is needed. Therefore, these kinds of solids are called high-energy solids, and their surfaces are called high-energy surfaces. Most molecular liquids can wet high-energy solid surfaces completely. Another kind of solid is held together by physical forces (van der Waals force and hydrogen bonds), which have a weak molecular lattice. Since such solids are unified through the weaker force, which can be broken by little energy, they are called low-energy solids, and their surfaces are called low-energy surfaces. According to the different kinds of liquids, some solid surfaces may be completely wetted while some others may be partially wetted (incomplete wetting) by liquids.

The wetting phenomenon is also a common phenomenon in nature. Many processes in animals and plants are closely related to the wetting phenomenon. Many production processes, such as cementing, lubrication, petrochemical, washing, welding, printing and dyeing, mineral flotation, waterproofing, and paint coating, are also closely related to the wetting phenomenon. Wetting theory is one of the theoretical bases for these production processes.

Therefore, research on wetting should not only focus on the phenomenon and the mechanism but also study its rules. Wetting is an essential adhesion of liquid on the solid surface, and more exactly, a competition between the adhesion force between liquid and solid surfaces and the cohesion among liquids themselves. Therefore, wetting involves the properties of liquid cohesion itself and adhesion with solid.

4.1.3 *Microscopic explanation of wetting phenomenon*

When two immiscible phases come into contact with each other, macroscopically, only one kind of wetting phenomenon is observed. However, its microcosmic interpretation needs to use interface layer theory. The so-called interface layer is a thin layer which is located near the interface (physical interface). According to the development of surface science in the present situation, there are three types of interface layer models. One is the Gibbs interface layer model with a segmentation surface type. Another is Guggenheim-type interface layer model with excessive strata. The third one is the interface layer model with a physical interface type. The physical interface type model affirms that there is a real physical interface phase between two phases. Both of the two contacted phases have their own interface sub-layers. The real physical interface was located between the two interface layers. The thickness of the interface sub-layer is roughly the same as the radius of molecular interaction. The changes in real physical properties of the interface on both sides are called mutation. Therefore, the changes in the properties of the interface layer between these two phases are mutation. The property of interface layer and that of inside bulk (body) are different. Whereas this difference is not mutation but a continuous process.

According to the physical interface layer model, the region at the juncture of liquid and solid materials is characterized by a thin liquid layer, the thickness of which corresponds to the radius of molecular interaction. This layer is referred to as the liquid interface that forms upon the contact between the liquid and solid phases. The

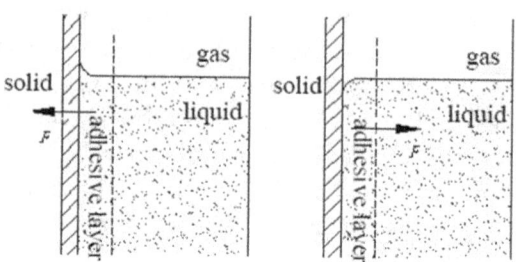

Figure 4.3.   Wetting and non-wetting.

liquid interface layer is shown by a dotted line in Figure 4.3. Macroscopically, it is called the "adhesive layer". The force acting on the molecules in the liquid interface layer, or "adhesive layer", is different from that inside the liquid bulk (body). The molecules in this layer, on the one side, are forced by solid-phase molecules lying on one side of the layer, which is the adhesion force, and on the other side, experience attractional force from the liquid body lying on the other side of the layer, which is the cohesion force. From the perspective of boundary layer, the side connected by the physical interface experiences the surface external force, and the side connected by the liquid experiences the surface internal force effect. In general, these two forces are not equal. Accordingly, the force acting on the molecules in this liquid interface layer is not symmetrical. When the cohesion force is greater than the adhesion force, the resultant force of the cohesion and adhesion forces of the liquid surface layer points in the direction toward the inside of the liquid, which tends to squeeze the molecules of the interfacial layer as much as possible inside the liquid. At this point of time, the solid is not wetted by the liquid. The liquid interface layer has the tendency of spontaneous contractions. The surface of the liquid has a convex shape. The contact angle for liquid–solid is an obtuse angle. When these phenomena appear, the solid is not wetted by the liquid. The essential reason for non-wetting is that the cohesion force is greater than the adhesion force. When the cohesion force is less than the adhesion force, the resultant force of the cohesion and adhesion forces of the liquid interface layer points in the direction away from the liquid (pointing toward the solid). The liquid internal molecules have the tendency to be pulled into the liquid interface layer. At this point of time, the solid is wetted by the liquid. The liquid interface layer has the tendency to spontaneously stretch.

The surface of the liquid has a concave shape. The contact angle for liquid–solid is an acute angle. When these phenomena appear, the solid is wetted by the liquid. The essential reason for wetting is that the adhesion force is greater than the cohesion force.

4.2 Influence Factors of Contact Angle

The size of the contact angle reflects the degree of interaction of liquid and solid surfaces and also the degree of wetting. If the attraction between the liquid and solid surfaces is very strong (such as water drops on a strongly hydrophilic solid), the liquid (drop) will totally spread on the solid surface. The contact angle is close to zero. If the solid is not strongly hydrophilic, the contact angle will reach 90°. In many strong hydrophilic solid surfaces, water droplets are found with contact angles ranging from 0° to 30°. If the solid surface is hydrophobic, the contact angle will be more than 90°. For superhydrophobic surfaces, the contact angle will reach 150° or even 180°. On these surfaces, water droplets will simply stay on the surface, without any substantial adhesion. These surfaces are called superhydrophobic surfaces, which can be obtained through fluorination (Teflon). The effect is also called "the lotus effect" because it is hydrophobic like a lotus, even rejecting sticky liquids such as honey. Table 4.2 shows the contact angles of different liquids and solids.

Table 4.2. Contact angle of different liquids and solids.

Liquid	Solid	Contact angle
Water Alcohol Diethyl ether Carbon tetrachloride Glycerol Acetic acid	Soda lime glass lead glass fused silica	0^0
Water	Solid paraffin	107°
	Silver	90°
Methyl iodide	Soda lime glass	29°
	Lead glass	30°
	Fused silica	33°
Mercury	Soda lime glass	14°

For different liquids and solids, the corresponding contact angle is different. Some of them have large differences. In addition, surface roughness, material composition, and temperature also have an influence on the contact angle.

4.2.1 *Influence on contact angle by surface roughness*

Surface roughness has a direct influence on the real contact angle. It can be described by the Wenzel model, which is shown as follows:

$$\cos \theta^* = \gamma_f \cos \theta \tag{4.1}$$

where θ^* is the apparent contact angle, θ is the ideal contact angle, which is also called the Young contact angle, and γ_f is the surface roughness ratio, which is defined as real solid superficial area divided by its apparent area.

4.2.2 *Influence on contact angle by the composition of surface materials*

The composition of surface materials has a direct influence on the contact angle. This relationship can be described by Cassie's law, which pertains to composite surfaces composed of two types, as shown in the following:

$$\cos\theta_c = f_1\cos\theta_1 + f_2\cos\theta_2 \tag{4.2}$$

where θ_c is the contact angle of the real composite material, θ_1 is the contact angle of material 1, f_1 is the proportion of material 1, θ_2 is the contact angle of material 2, and f_2 is the proportion of material 2, $f_1 + f_2 = 1$.

For various materials, equation (4.2) can be expanded as follows:

$$\cos\theta_c = \sum_{i=1}^{N} f_i\cos\theta_i \tag{4.3}$$

and $\sum f_i = 1$.

4.2.3 *Contact angle changing with temperature*

Contact angle is associated with the degree of wetting, which reflects the molecular cohesion and the adhesive forces between molecules at the interface. Variations in temperature can influence molecular movement, spacing, and intermolecular forces. Consequently, the contact angle is closely linked to temperature.

4.3 Contact Angle Hysteresis and Influencing Factors

When liquid drops fall on a non-ideal real solid surface, the droplets will unfold from their original spherical shape to a spherical crown on the solid surface. Then, a contact angle is formed. Since this stable state unfolded from a drop, the contact angle is called the advancing contact angle, which is represented by θ_a. If we use a method to pack up the spherical crown from the solid surface, at the beginning, the three-phase contact wire does not move, and only the contact angle reduces. When the contact angle has reduced to a certain degree, the three-phase contact wire starts to shrink. This change in contact angle from the process of taking back is called the receding contact angle, which is represented by θ_r. For a real solid surface, the receding contact angle is less than the advancing contact angle.

The phenomenon of the receding contact angle being less than the advancing contact angle is called contact angle hysteresis. The difference between these two contact angles is the hysteresis contact angle. For an ideal solid surface, the receding contact angle is equal to the advancing contact angle.

There are many factors that influence contact angle hysteresis. The most important factor is solid surface roughness (or non-smoothness). Since a real surface is not very smooth, the liquid–gas surface is suffocated during expansion. It is possible to further expand only after attaining a maximum angle. Moreover, since a real surface is not very smooth, receding is suffocated. It can be taken further back only after attaining a minimum angle. Therefore, the surface roughness has the most direct influence on contact angle hysteresis. In addition, the presence of pollution on surfaces, the accumulation of sediment, and three-phase line movement speed also have great effects on contact angle hysteresis.

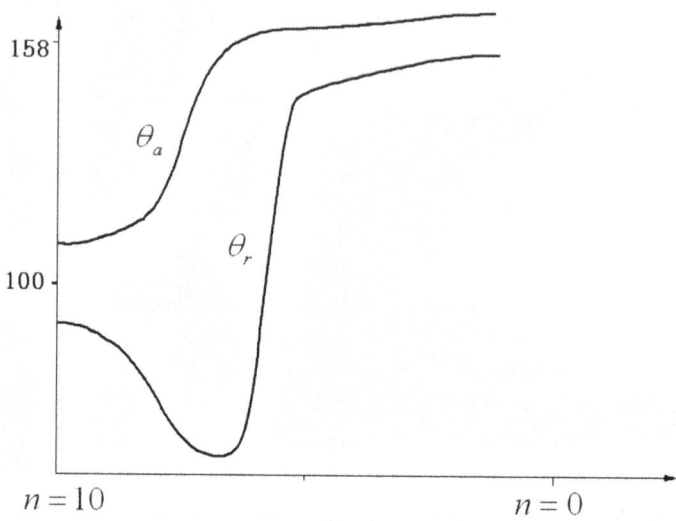

Figure 4.4. Advancing and receding contact angles for water on PTFE wax.

4.3.1 *Influence of surface roughness*

Dettre and Johnson obtained a set of classic data for water on polyte-trafluoroethylene (the PTFE, or Teflon) wax blends through a series of experiments, as shown in Figure 4.4. They used spray wax to produce a rough surface and made the surface smoother by furnace heating. In the graph, the abscissa n expresses the number of heat treatments. The roughness decreases with an increase in the number of heat treatments. The change in the receding contact angle θ_r with the roughness is nonmonotonic.

4.3.2 *Influence of surface pollution*

Pollution on the surface of a liquid or solid can easily cause the hysteresis phenomenon. For example, if a solid plate surface has grease, upon slowly submerging it into water, the water will be blocked by the polluted locations. Then, it forms a large contact angle. When water spills over grease, most oil will unfold on the surface of water. When the receding contact angle is measured by revealing the solid from the water, lower values will be obtained. Conducting experiments by using graphite and talc, Fowkes and Harklns found that the contact angle hysteresis phenomenon can in fact be eliminated by strictly purifying the liquid and solid.

4.3.3 *Influence of deposits on solid surface*

Solute in liquid, such as surfactants and polyphosphate content, can deposit a layer of film on a solid surface. In some cases, the existence of this film can cause the hysteresis effect. Generally, once the film is formed, it will stick on the surface.

4.4 Mechanics Balance Relationship of Three-Phase Line and the Young Equation

When a certain kind of incompatible liquid is dropped on a liquid surface, it presents three phases (gas α, liquid β, and liquid θ) as illustrated in Figure 4.5. There is a triple-phase line at the triple-phase meeting point. In equilibrium, the unit-length resultant force of the three-phase boundary should be zero.

Their projection components in any three-phase border direction should also be zero:

$$\gamma_{\alpha\beta} + \gamma_{\theta\beta}\cos\theta + \gamma_{\alpha\beta}\cos\alpha = 0 \tag{4.4}$$

$$\gamma_{\alpha\theta}\cos\theta + \gamma_{\theta\beta} + \gamma_{\alpha\beta}\cos\beta = 0 \tag{4.5}$$

$$\gamma_{\alpha\theta}\cos\alpha + \gamma_{\theta\beta}\cos\beta + \gamma_{\alpha\beta} = 0 \tag{4.6}$$

where α, β, and γ each represent a phase and the angle of this phase. γ_{ij} is the surface free energy (interface free energy) or surface tension (interface tension) between the two phases. This relationship can be shown by a similar triangle, as seen in Figure 4.6.

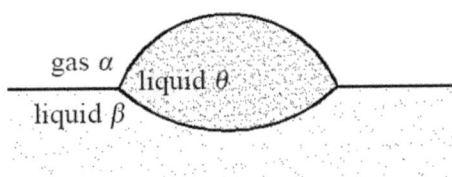

Figure 4.5. Three-phase line.

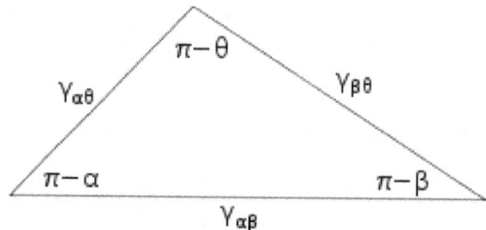

Figure 4.6.   Neumann triangle.

The triangle is called the Neumann triangle, which meets the following geometric condition:

$$\alpha + \beta + \theta = 2\pi \tag{4.7}$$

This geometry relationship can be used to deduce the surface free energy of the three-phase relationship. From this triangle, the sum of any two sides is greater than the third side, $\gamma_{ij} < \gamma_{jk} + \gamma_{ik}$. This indicates that no one energy can exceed the total of other interface free energies.

If phase β is a rigid smooth surface, $\beta = \pi$. The above three equations can become a single equation:

$$\gamma_{SV} = \gamma_{SL} + \gamma_{LV}\cos\theta \tag{4.8}$$

where S is the rigid smooth solid surface phase, L is liquid, and V is gas. This equation is known as the Young equation. From this equation, we can understand that if the surface free energies of the three phases are known, the equation can be used to determine the contact angle. Furthermore, if one of the gas phases is replaced by a different type of liquid, this equation remains applicable.

4.5   Types of Wetting

Generally, the wetting phenomenon is categorized into three types: bedewing, soaking, and spreading. These three types of wetting represent distinct methods of application. One of the most common forms of wetting is spreading.

4.5.1 *Bedew*

The essence of bedewing is that a liquid adheres to a solid surface. The process of bedewing means that when the liquid and solid touch each other, a liquid–gas interface and a solid–gas interface are changed into a solid–liquid interface.

When two phases touch each other, the intermolecular forces that exist between them are referred to as "adhesion". Therefore, the essence is that the liquid adheres to the solid surface.

As shown in Figure 4.7, phase A and phase B are defined, where phase A represents the liquid and phase B represents the solid phase. The adhesion process occurs when phase A and phase B come into contact with each other; the contact surface between phase A and gas and the contact surface between phase B and gas are replaced by the contact surface between phase A and phase B.

Before adhesion, the free surface energy of phase A is γ_A and that of phase B is γ_B. After adhesion, these two energies will disappear and be replaced by the interfacial free energy between A and B. The contact area between A and B is supposed to be a united area. Under constant temperature and pressure, the change in system free energy caused by the adhesion process is

$$\Delta G = \gamma_{AB} - \gamma_A - \gamma_B \qquad (4.9)$$

where γ_{AB}, γ_A, and γ_B, respectively represent interfacial free energies for the united area of A and B, A and gas, and B and gas.

According to the work–energy conservation, reduction of the free energy is mainly used in the two-phase adhesion effect. Therefore, the reduced value should be equal to the increased value of adhesion work. Therefore, in terms of absolute value, the work of adhesion

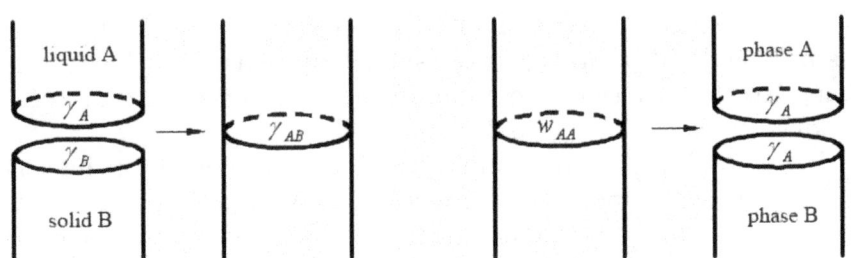

Figure 4.7.   Adhesion work and cohesion work.

W_{AB} is equal to the change in the system free energy but with an opposite sign:

$$W_{\mathbf{AB}} = -\Delta G = \gamma_{\mathbf{A}} + \gamma_{\mathbf{B}} - \gamma_{\mathbf{AB}} \qquad (4.10)$$

This equation verifies that if $W_{AB} \geq 0$, $\Delta G \leq 0$. The adhesion process (i.e., the bedew process) can be spontaneous. The smaller the common interfacial tension between A and B, the greater the W_{AB}, the easier the adhesion process. Thus, the solid is bedewed more easily by the liquid.

Another implication for adhesion work W_{AB} is that it is the work that is used to separate the unit area of phase A and phase B interface into the interfaces of phase A and phase B with gas. Obviously, the greater the W_{AB}, the more firmly A and B will contact and combine with each other, and the stronger the molecular interactions between A and B. Therefore, the value of the adhesion work W_{AB} reflects the magnitude of molecular interaction between two phases.

In a special case, if the discussed two phases are the same phase, such as phase A, the work used to separate A is not adhesion work but cohesion work. The cohesion work for A is

$$w_{\mathbf{AA}} = \gamma_{\mathbf{A}} + \gamma_{\mathbf{A}} - 0 = 2\gamma_{\mathbf{A}} \qquad (4.11)$$

The cohesion work W_{AA} is equal to the work required for opening and separating cylinder A, as shown in Figure 4.7. The same cohesion work reflects not only the strength of the combination of phase A itself but also the magnitude of molecular interaction of phase A.

4.5.2   *Soak*

When the clean surface of a solid is immersed in liquid, the solid–gas interface will be changed to a solid–liquid interface. There is no change to the surface of the liquid in such a process, which is called soaking, as shown in Figure 4.8.

If the total area of the solid block is a unit area, since the solid surface free energy before the immersion is γ_{B}, and the original solid surface is changed to a liquid contact interface after immersing, its interface free energy is γ_{AB}, and the liquid surface free energy has no change. Therefore, under constant temperature and pressure, the

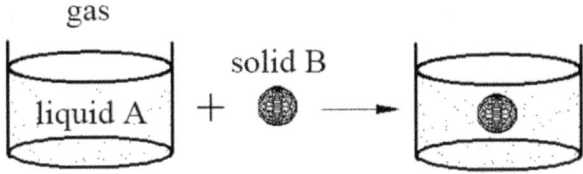

Figure 4.8. Soaking process.

change in system free energy caused by the process is as follows:

$$\Delta G = \gamma_{\mathbf{AB}} - \gamma_{\mathbf{B}} \qquad (4.12)$$

The work done by the process is called soak work:

$$W_{\mathbf{A/B}} = -\Delta G = \gamma_{\mathbf{B}} - \gamma_{\mathbf{AB}} \qquad (4.13)$$

According to the equation above and superficial phenomenon theory, if soak work $W_{\mathbf{A/B}} \geq 0$, $\Delta G \leq 0$. The process can be spontaneous. For the bedewing process mentioned above, it has been explained that, since the solid–liquid common interfacial tension is always less than the sum of their respective surface tensions, the adhesion work is greater than zero, so bedewing process can be spontaneous. However, soak is different. Only when solid surface free energy is greater than the solid–liquid common interface free energy, $W_{A/B}$ is greater than zero. Only in this case can the soak process occur spontaneously. This means that not all liquids and solids can be soaked spontaneously.

4.5.3 *Spreading*

Bedew and soak are special cases. Normally, liquid spreads after bedewing solid. The soak process always goes through the liquid surface. Therefore, common soak is always accompanied by spreading. The Neumann triangle and the Young equation predict that any one of solid–gas surface free energy $\gamma_{\mathbf{SV}}$, the liquid–solid interface free energy $\gamma_{\mathbf{SL}}$, and liquid gas–interface free energy $\gamma_{\mathbf{LV}}$ cannot be more than the sum of the free energy of the other two surfaces. The restrictions are on situations for partial wetting. By breaking this restriction, the conditions can be determined for the complete wetting or total non-wetting. If $\gamma_{\mathbf{SV}} > \gamma_{\mathbf{SL}} + \gamma_{\mathbf{LV}}$, it can be considered as complete wetting. When $\gamma_{\mathbf{SL}} > \gamma_{\mathbf{SV}} + \gamma_{\mathbf{LV}}$, it can be considered as total non-wetting. In this case, the Young equation has no physical solution, which means the contact angle does not have a balance state between 0° and 180°.

The spreading parameter S is a useful parameter to measure wetting, which is defined as

$$S = \gamma_{\textbf{SV}} - (\gamma_{\textbf{SL}} + \gamma_{\textbf{LV}}) \tag{4.14}$$

When $S > 0$, the liquid completely wets the solid surface. And when $S < 0$, it only partly wets. Substituting the Young equation into (4.14) leads to

$$S = \gamma_{\textbf{LV}}(\cos\theta - 1) \tag{4.15}$$

This equation is called the Young–Dupre equation, which has a physical solution only when $S < 0$.

This is similar to the discussion on the contact angle. Some quantitative liquid A is put on the surface of solid B or another liquid. Then, there are two cases: one is that the liquid would form drops on the discussed surface or a lens on another liquid; the other is that the liquid would spread to form a liquid film on the discussed surface or a thin film on another liquid.

Both these conditions can be called wetting phenomena. However, their wetting degrees are different. A contact angle of $\theta < 90°$ usually indicates partial wetting of the solid surface, which is called partial wetting. $\theta = 0°$ means "perfect wetting", and the liquid will spread over the surface. Contact angles greater than 90° generally mean non-wetting to the solid surface, as shown in Figure 4.9.

When $\theta \leq 0°$, liquid droplets can automatically spread over the solid surface to form a liquid film. The liquid droplets can spread over the solid surface.

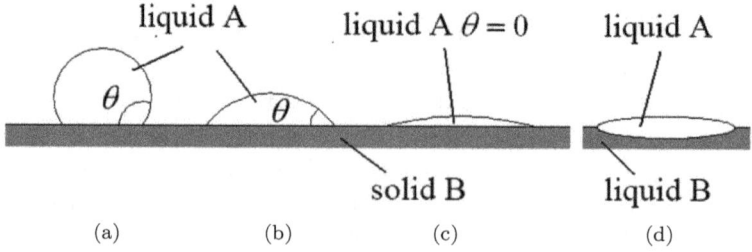

(a)   (b)   (c)   (d)

Figure 4.9.   Spread condition of liquid on liquid or solid. (a) Wetting without spread, (b) wetting, (c) perfect wetting spread, and (d) liquid spread to another liquid.

During spreading, the solid–gas interface disappears and the solid–liquid and liquid–gas interfaces are formed. Based on energy change, under constant temperature and pressure, the free energy change of system is

$$dG = \left(\frac{\partial G}{\partial A_{LV}}\right) dA_{LV} + \left(\frac{\partial G}{\partial A_{LS}}\right) dA_{LS} + \left(\frac{\partial G}{\partial A_{SV}}\right) dA_{SV} \quad (4.16)$$

where dA is an area variable and the subscripts L, S, and V represent liquid, solid, and gas, respectively. During the spreading process, the increased area of the liquid–gas is equal to the decreased area of the solid–gas interface. Hence,

$$dA_{LV} = -dA_{SV} = dA_{LS} \quad (4.17)$$

Since

$$(\partial G/\partial A_{LV}) = \gamma_{LV}; (\partial G/\partial A_{LS}) = \gamma_{LS}; \quad (\partial G/\partial A_{SV}) = \gamma_{SV} \quad (4.18)$$

we have

$$\frac{dG}{dA_{LV}} = \left(\frac{\partial G}{\partial A_{LV}}\right) + \left(\frac{\partial G}{\partial A_{LS}}\right) - \left(\frac{\partial G}{\partial A_{SV}}\right) = \gamma_{LV} + \gamma_{LS} - \gamma_{SV} \quad (4.19)$$

By defining

$$S = -\left(\frac{dG}{dA_{LV}}\right) \quad (4.20)$$

where S is the spreading coefficient of the liquid over the solid surface, we have

$$S = \gamma_{SV} - \gamma_{LV} - \gamma_{LS} \quad (4.21)$$

It is the change in free energy of the system during liquid spreading over the solid surface. It is also the same as the previously defined spreading coefficient. From the perspective of energy, it has a new meaning: the spreading coefficient is the negative energy change which is caused by increasing unit liquid surface or solid–liquid surface. If S is positive, it indicates that the system free energy can decrease and the spreading is spontaneous. If S is negative, it indicates that the system free energy needs to increase and the spreading is not spontaneous, which can only form liquid droplets on the surface. Hence, from the value of S, we can judge the ability of spreading over the solid.

Generally, organic liquids can spread over high-energy surfaces (metals and glass), but they cannot spread over low-energy liquids (paraffin and polyethylene), particularly when surface tension is very high. Presently, the most difficult to spread is the solid which has — CF_3, secondly $-CF_2-$, or $-CH_3$ and $-CF_2-$. The practical situation reflects that the spreading ability of the discussed surface relates to not only the cohesion force of molecular interactions but also the adhesion forces of molecular interactions between the liquid and solid surfaces.

Since the spreading process is an unbalanced process of the system but the spreading coefficient above is based on energy or surface free energy for the balance system, it is called traditional spreading coefficient.

The spreading coefficient can be rewritten as

$$S = \gamma_{SV} - \gamma_{LV} - \gamma_{SL} = \gamma_{SV} + \gamma_{LV} - \gamma_{SL} - 2\gamma_{LV} = W_{SL} - W_{LL}$$

(4.22)

where W_{SL} is the solid–liquid adhesion work and W_{LL} is the liquid cohesion work. When the adhesion work between the liquid and the solid is greater than the liquid cohesion, $W_{SL} > W_{LL}$, $S > 0$, the liquid can spread over the solid. When the adhesion work between the liquid and the solid is less than the liquid cohesion, $W_{SL} < W_{LL}$, $S < 0$, the liquid cannot spread over the solid.

4.6 Principle and Method of Measuring Contact Angle

There are many measurement methods for the contact angle, including both direct and indirect methods. By directly taking photos of the shape of droplets or bubbles that are spreading on a solid surface, the contact angle can be measured by assessing the angle and height from the photos. This measurement is a direct method. Other direct methods also include the use of specular reflection law or parallel beams to measure the contact angle. Calculating the contact angle from the capillary rise or decline in height by capillarity is an indirect method. Indirect measurement methods also include the force method, among others, for measuring contact angle. Currently, there are many kinds of equipment for measuring the contact angle. Some commonly used methods and principles of contact angle measurement are introduced here.

4.6.1 The principle of measurement for contact angle by hypsometry

When the volume of a drop is less than $6\,\mu\text{L}$, the influence of gravity on its shape can be negligible, and the graph of the drop viewed from the front is thought to be part of a standard circle. As shown in Figure 4.10, if the height h of the liquid on the solid surface and the diameter D of the solid contact area can be measured, the contact angle θ can be calculated. According to a geometrical relationship, the computational formula can be written as

$$\theta = 2\arctan\frac{2h}{D} \tag{4.23}$$

where θ is the contact angle, h is the height of the liquid spherical crown, and D is the diameter of the bottom circle of the spherical crown.

Whether $\theta > 90°$ or $\theta < 90°$, this equation is applicative. This equation is deduced as follows.

For right-angled triangle ACO', $\alpha + \beta = 90°$, then

$$\theta = 90° + (\beta - \alpha) = (\beta + \alpha) + (\beta - \alpha) = 2\beta \tag{4.24}$$

$\tan\beta = \frac{h}{r}$, so, $\beta = \arctan\frac{h}{r} = \arctan\frac{2h}{D}$

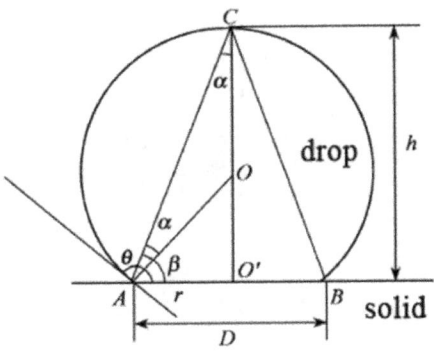

Figure 4.10. Hypsometry method to measure the contact angle.

Then,

$$\theta = 2 \arctan \frac{2h}{D} \qquad (4.25)$$

4.6.2   *Photo goniometric method*

The tool used in this method is an isosceles right-angled protractor. The two sides of the isosceles right-angled protractor are, respectively, a and b. The protractor can be moved up and down or left and right. Its measurement process is as follows:

1. Move down the a and b sides until they are tangent to the liquid, as shown in Figure 4.11.
2. Continue moving down verticality until the peak of the protractor and the border of drop are crossed at point C. Therefore, the coordinate of the peak C can be defined, as shown in Figure 4.12.
3. Make the protractor counterclockwise and rotate around point C; turn δ angle until the side A intersects with the crossed point A' of the three phase surfaces, as shown in Figure 4.13. Then, θ can be obtained.

According to a geometrical relationship,

$$\theta = 90° + (\beta - \alpha) = (\beta + \alpha) + (\beta - \alpha) = 2\beta$$

Since $\beta = 45°+\delta$, $\theta = 90°+2\delta$. When the contact angle $\theta > 90°$, ide AC of the protractor is rotated counterclockwise and δ is positive; when the contact angle $\theta < 90°$, side AC of the protractor is rotated clockwise and δ is negative.

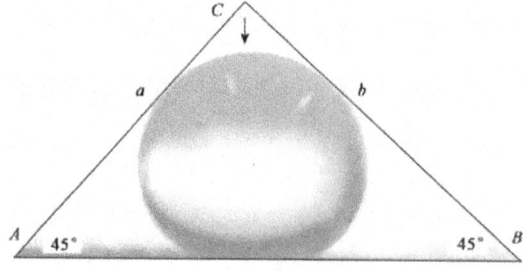

Figure 4.11.   Tangency with both sides of protractor figure.

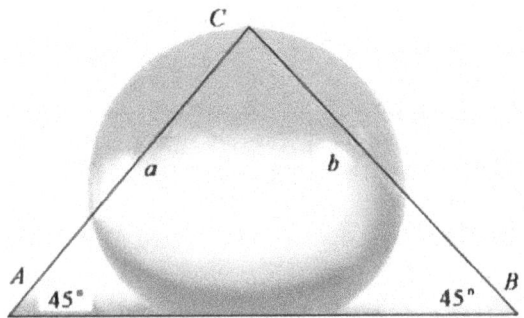

Figure 4.12. Parallel moving downward and made peak coincidence with peak of sphere.

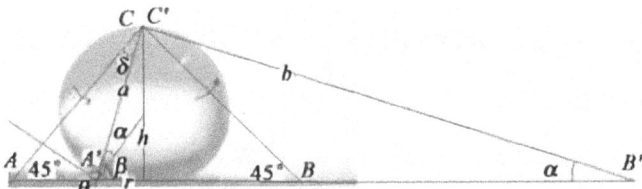

Figure 4.13. Left-handed rotation protractor and make solid–liquid–gas crossing.

4.6.3 *Specular reflection method*

Langmuir and Schaeffer put forward through droplets a mirror to measure the contact angle. As shown in Figure 4.14, a single beam shines at the three-phase contact line. When and only when the incident angle is equal to the contact angle, the observer can see reflected light in the incident direction. Fort and Patterson improved this method, developed it into the production, and made its accuracy reach ±1°. The actual device diagram is shown in Figure 4.15. The end of a lever which can rotate around the fixed point is equipped with an illuminant, which can produce a single beam. At the point where it is close to the illuminant, there is a peephole which is used to observe the reflected light. Rotation of the lever is realized through screw motion. The key to this operation is to make the bottom of the tested droplets and the rotation axis of the lever coplanar. This can guarantee that the single beam transmits to the contact line of the droplet, which has a certain difficulty in actual practice. Good and Ferry measured the contact angle of stainless steel in liquid hydrogen

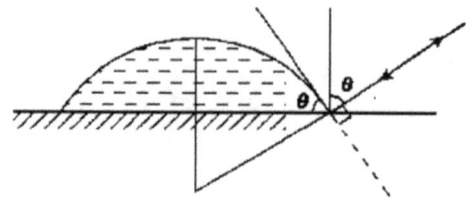

Figure 4.14. Schematic diagram of reflected light of a single beam.

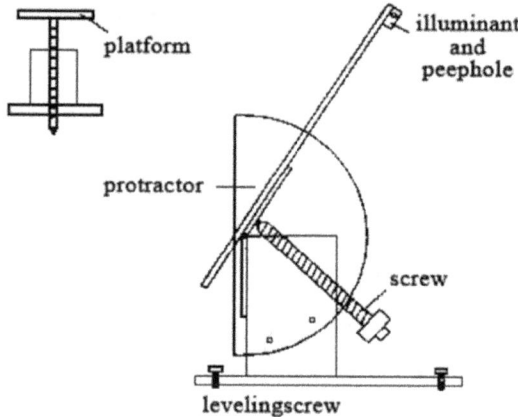

Figure 4.15. Tester of mirror reflection.

by uniting the tilting plate method and the reflection method and made its accuracy reach ±0.5°. There are many advantages to this method, such as avoiding the use of a complex optics system, simple equipment, low cost, and ease of processing. However, it can only measure contact angles which are less than 90°.

4.6.4 *Parallel beam method*

Allain's parallel light method is essentially a further development of the method of specular reflection, as shown in Figure 4.16. Parallel beams entering from right above the droplet are reflected by the gas–liquid interface around the droplet and form two angles with the incident direction. A screen is placed at a point that has a distance s from the measured surface. The image of the bottom of the amplifying droplet appears on the screen. Its magnification depends on s and the contact angle. The simple geometrical relationship shows

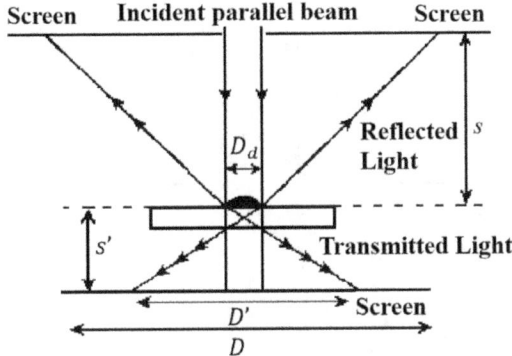

Figure 4.16. Light path chart of the parallel beam method.

that

$$\theta = \frac{1}{2} \arctan \left(\frac{D - D_d}{2s} \right) \tag{4.26}$$

This method can measure contact angles which are less than 45° and the accuracy can reach +0.1°.

For surfaces capable of transmitting light, such as glass, Allain proposed the parallel light transmission method. Parallel beams entering right above are refracted by the gas–liquid interface around the contact wire, enter inside the droplet, and are then refracted once again by the liquid–solid interface into the gas phase below. Two refracted beams and the incident direction form θ'' angle. If the screen below is placed at the point which has a distance s' from the measured surface, the inverted image of the bottom of the amplifying droplet also appears on the screen.

4.6.5 Calculating contact angle from rise and descent method by capillary tube

As shown in Figure 4.17, R_1 is the radius of the supposed capillary tube and R is the radius of curvature of the meniscus formed by the liquid in a tube. Then, the contact angle θ can be calculated using equation (4.27):

$$\cos \theta = \frac{R_1}{R} \tag{4.27}$$

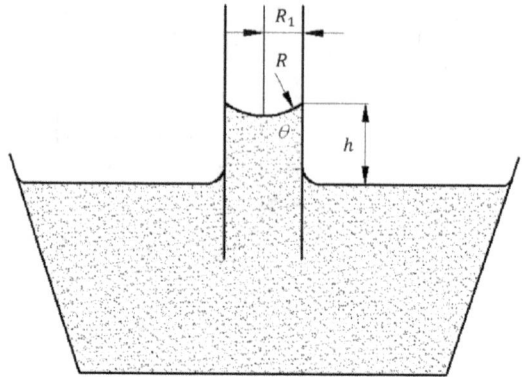

Figure 4.17. Capillary rising.

R_1 can be measured from the pipe diameter and R can be obtained from the following:

$$\Delta\rho \cdot gh = \frac{2\gamma}{R} \tag{4.28}$$

$\Delta\rho$ and h can be measured, and γ can be calculated from another measurement.

Different measuring methods correspond to different principles, and the quality of the measuring method depends on the accuracy of measurement. With the unceasing development of optical measurement technology, the measurement precision of contact angle is also gradually improving. Since the contact angle is closely related to surface tension, many measurement methods of contact angle are accompanied by the measurement of surface tension. Some of the methods above can also be used to measure the contact angle, such as the dynamometry method.

4.6.6 *Principle of dynamometry to measure contact angle*

This method is also called Wilhemly (broadly) or Wilhemly type and was suggested by Wilhemly in 1863. The device is shown in Figure 4.18. A test solid sheet is connected to an electronic balance through a wire. When the sheet is not dipped into the liquid, it is

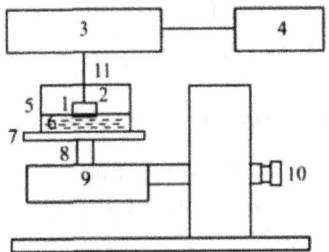

Figure 4.18. Wilhemly device schematic diagram. (1) tested solidsheet, (2) wire, (3) electronic balance, (4) recorder, (5)tested unit, (6) tested liquid, (7) removable platform, (8) hoistable platform, (9) electrical machine, (10) support, (11) cup.

only supported by gravity, as indicated by the force on the device:

$$F_1 = mg \tag{4.29}$$

where F_1 is the force when the sheet has not been dipped into the liquid, m is the mass of the tested solid sheet, and g is acceleration due to gravity.

When the sheet is dipped into a depth of h, it reaches a balance:

$$F_2 = mg + c\gamma \cos\theta - \Delta\rho g V_{\text{disp}} \tag{4.30}$$

This represents the device when the depth is h. c is the wetting perimeter, γ is the surface tension of the liquid, θ is the contact angle, $\Delta\rho$ is the density difference between liquid and gas, h is the depth of immersion, and V_{disp} is the volume of the immersed liquid.

The difference in the force measurement device before and after immersing the solid sheet into liquid is

$$\Delta F = F - F = c\gamma \cos\theta - \Delta\rho g V_{\text{disp}} \tag{4.31}$$

The contact angle can be calculated by the measurement method. The emergence of electronic balance greatly improved the accuracy, which changed from the early $\pm 1°$ to $\pm 0.1°$. Dynamic surface energy analyzers of type K12, produced by a German company, are designed based on the Wilhemly principle.

4.7 Measuring Method of Contact Angle Tester Based on Photogoniometric Method

4.7.1 *Construction compose of JYSP-180 tester*

The instrument consists of an illuminant, worktable, base, magnifying glass, CCD camera, and liquid dropping device, as shown in Figure 4.19.

4.7.2 *Installation and use of tester*

1. Install and adjust

A. Put the tester on a stable platform. Room temperature should be constant at $20 \pm 5°C$. Adjust the leveling knob and make the tester horizontal based on a level bubble.

B. Connect the magnifying glass to the CCD camera, and then install them on lifting bearings.

C. Connect the CCD camera and the video capture card using a video cable. Connect the CCD camera power.

D. Insert the power plug in the external socket which has reliable grounding. Turn on the power switch.

E. Adjust the working condition of the tester, as follows.

Enter the video collection window. By adjusting the workbench, lift the bearing and magnifying glass tube. The objective table will appear in the video window. Adjust the image until it is clear. Unlock the locking button. Turn on the camera and align the upper edge

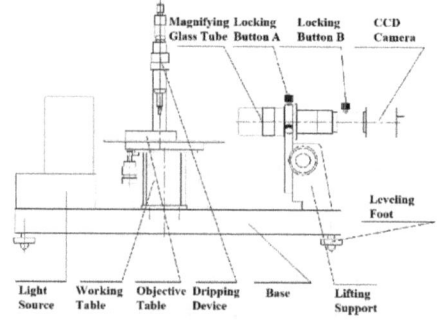

Figure 4.19. Structure of contact angle tester.

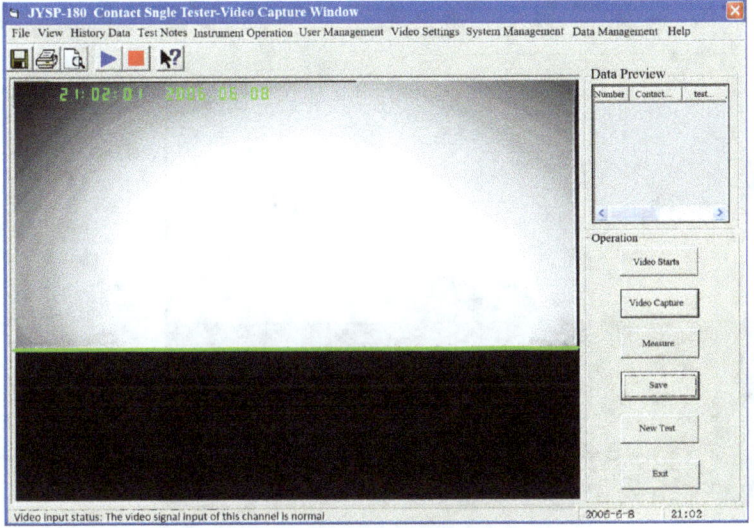

Figure 4.20. Initialized states.

of the objective table with the green line in the window, as shown in Figure 4.20. Then, tightly lock the lock button to complete the adjustment of the equipment.

2. Use tester

A. Place the solid sample in the small objective table.
B. Utilize a dispensing apparatus to administer an appropriate volume of liquid onto a solid sample.
C. Adjust the workbench, lift bearing, and magnifying glass tube so that the image is clear.
D. Press the "video capture" to record clear images and continue to see if there are clearer images. If there is any, continue to press the "video capture" to record the clear images. Click "measured" and choose the measurement method. If a clearer image is needed, directly click "measurement" and choose different measurement methods.

3. Operating steps

A. As shown in Figure 4.21, move the measuring ruler up and down or left and right

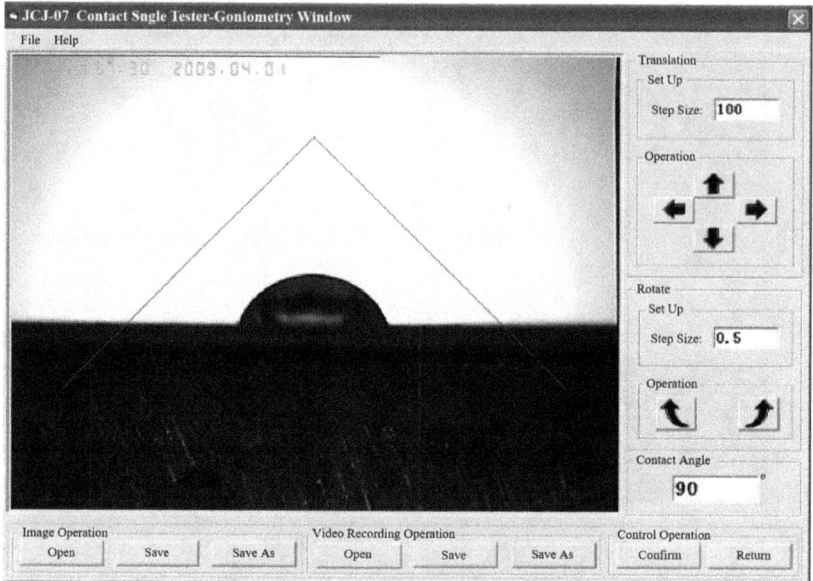

Figure 4.21.   Move the ruler.

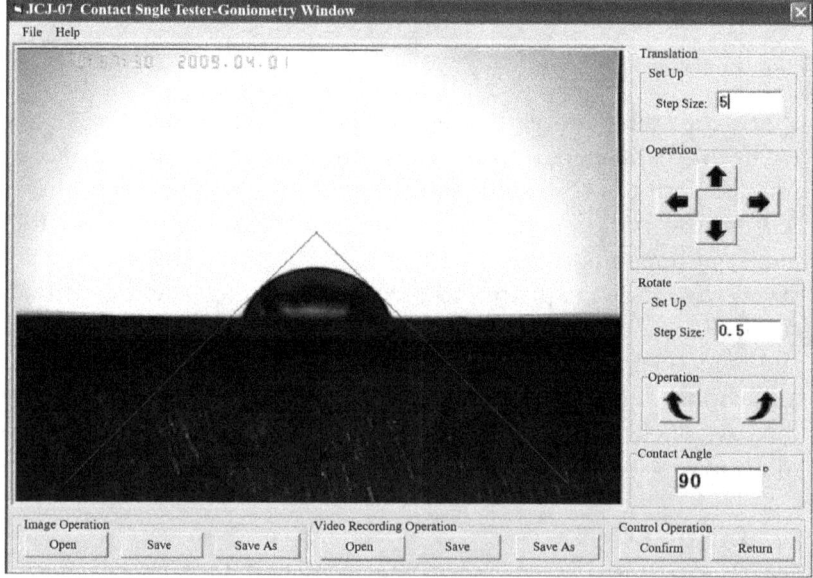

Figure 4.22.   Ruler and droplet edge tangent.

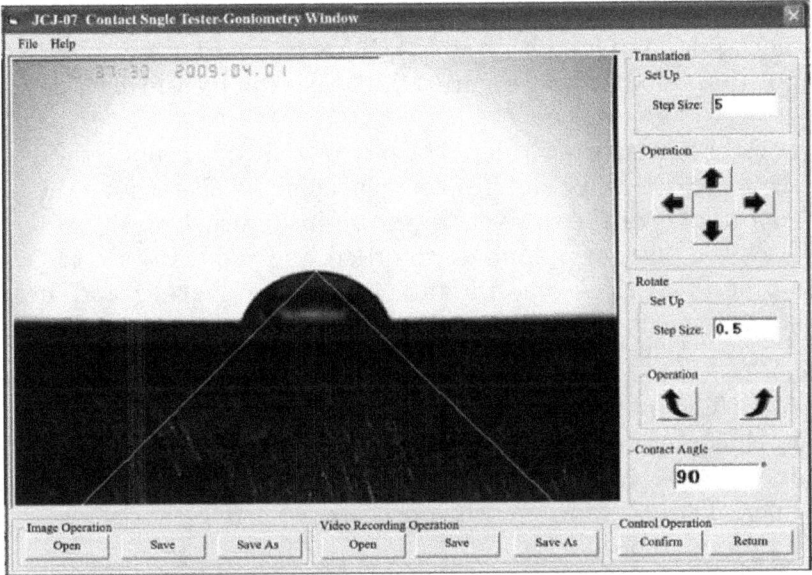

Figure 4.23. Demarcate the high point of the droplet by ruler.

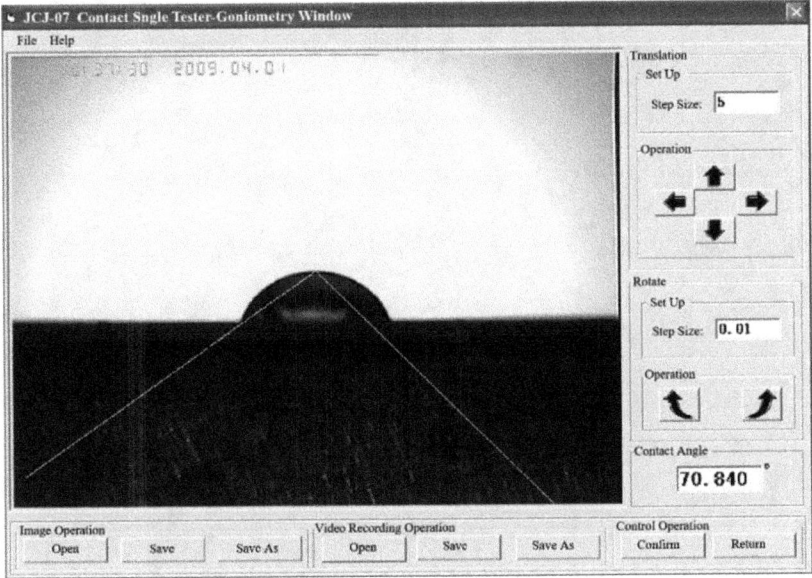

Figure 4.24. Rotate ruler to find contact angle.

Note: this time, only move horizontally or vertically the measurement ruler and do not rotate it). Move the measuring ruler to the position, as shown in Figure 4.22. Make the measuring ruler and droplet edge tangent.

B. Lock the measuring ruler. Make it overall down until the intersection of the ruler and droplet edge coincide. Determine the high coordinates of the droplet, as shown in Figure 4.23.

C. Lock the buttons up and down or left and right. Rotate the measuring ruler, as shown in Figure 4.24. Make the ruler and droplet side intersect, and then click the "ok" button to complete the contact angle measurement.

Note: The translation and rotation test process can be accelerated by changing the step length: directly use the "up", "down", "left", and "right" arrow keys on the keyboard to move the measuring ruler. The "Z" and "Y" buttons are for left rotation and right rotation, respectively.

D. Click "return" to continue with the next time.

Chapter 5

Young–Laplace Equation and Its Application

5.1 Young–Laplace Equation

The most typical example of capillarity phenomena is the rise and fall of liquid in a capillary tube, which results from the surface tension of the liquid. Its motivation comes from the pressure difference between the high and low layers of liquid in the capillary tube. The curved surface of the high layer of the liquid is related to both surface tension and pressure difference. What is the relationship among pressure, surface tension, and curved surface? In order to discuss such a relationship among the pressure on both sides of the curved surface, the liquid surface tension, and the geometric shape of the curved surface, a micro-curved surface unit 1234 is considered. The micro-surface is cut by two orthogonal normal incisions. The so-called normal incision is the intersection line of the plane containing normal lines and the curved surface. If two normal incisions are in the same directions, respectively, as the maximum and minimum curvature directions, the curvatures of curves 12 and 14 are the two principal curvatures of the curved surface unit.

As shown in Figure 5.1, for the micro-unit 1234, curvature radius (principal curvature) of curves 12 and 34 is r_1, the arc center angle is $d\phi_1$, and the arc length is a. The curvature radius (another principal curvature) of curves 14 and 23 is r_2, the arc center angle is $d\phi_2$, and the arc length is b. Then the area of the micro-unit is $dA = a \cdot b = r_1 \cdot d\phi_1 \cdot r_2 \cdot d\phi_2$. The surface tension of the curve is supposed to be γ,

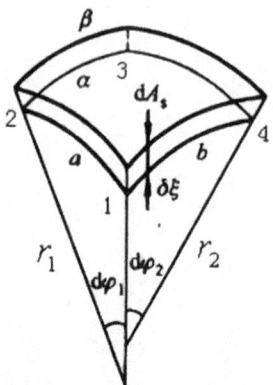

Figure 5.1.   Micro-volume and superficial area.

and the difference in pressures acting in the normal direction on both sides of the micro-unit is Δp (the pressure difference between the concave side and the convex side of the curve surface). Thus, when the curve surface has a virtual displacement $\delta\xi$ in the normal direction, the increased virtual area of the micro-unit (relative to the original area) is

$$\delta dA = \delta\xi \cdot d\phi_1 \cdot b + \delta\xi \cdot d\phi_2 \cdot a$$
$$= \delta\xi(r_2 d\phi_2 \cdot d\phi_1 + r_1 d\phi_1 \cdot d\phi_2) \qquad (5.1)$$
$$= \delta\xi(r_1 + r_2)d\phi_1 \cdot d\phi_2$$

Under the effect of pressure difference Δp, when virtual displacement $\delta\xi$ is produced, virtual work is $\delta W = \Delta p \cdot dA \cdot \delta\xi$. Due to the increased virtual area (overcoming the surface tension), the increased virtual energy on the surface is $\delta E = \delta dA \cdot \gamma$. Based on the conservation between work and energy, in the case of no other external energy, there is $\delta W = \delta E$, so that

$$\Delta p \cdot r_1 r_2 d\phi_1 d\phi_2 \delta\xi = \delta\xi(r_1 + r_2)d\phi_1 \cdot d\phi_2 \qquad (5.2)$$

The solution is obtained as

$$\Delta p = \gamma\left(\frac{1}{r_1} + \frac{1}{r_2}\right) \qquad (5.3)$$

This is the Young–Laplace equation, which was deduced independently by T. Young and P.S. Laplace in 1805 and 1806, respectively. Actually, the Young–Laplace equation can also be deduced by mechanical equilibrium. Similarly, the micro-curved surface unit is considered. The tension acted on the surface is supposed to be γ, the pressure acted on the concave side of surface is P_{concave}. The convex side of surface is P_{convex}, then pressure difference is $\Delta p = P_{\text{concave}} - P_{\text{convex}}$. This pressure can be balanced by the resultant force of components in normal direction of surface tension. Components in tangential direction of γ can be balanced by itself. The resultant force in normal direction is

$$f_s = \gamma \cdot b \cdot \sin \frac{d\phi_1}{2} \times 2 + \gamma \cdot a \cdot \sin \frac{d\phi_2}{2} \times 2 = \gamma \cdot r_2 d\phi_2 \cdot \sin \frac{d\phi_1}{2} \times 2$$

$$+ \gamma \cdot r_1 d\phi_1 \sin \frac{d\phi_2}{2} \times 2 \tag{5.4}$$

When the micro-unit is infinitely tiny, $\sin(d\phi_1/2) = (d\phi_1/2)$, $\sin(d\phi_2/2) = (d\phi_2/2)$,

$$f_{\mathbf{s}} = \gamma(r_1 + r_2)d\phi_1 d\phi_2 \tag{5.5}$$

The resultant force of pressure difference Δp in normal direction is

$$f_\gamma = \Delta p \cdot a \cdot b = \Delta p \cdot r_1 d\phi_1 \cdot r_2 d\phi_2 \tag{5.6}$$

Under equilibrium condition, $f_{\mathbf{s}} = f_\gamma$, i.e.,

$$\gamma(r_1 + r_2)d\phi_1 \cdot d\phi_2 = \Delta p \cdot r_1 r_2 d\phi_1 d\phi_2 \tag{5.7}$$

The solution is

$$\Delta p = \gamma \left(\frac{1}{r_1} + \frac{1}{r_2} \right) \tag{5.8}$$

If the two sides of the surface are in two different phases, the pressure of convex phase should be greater than the pressure of concave phase. If convex phase is liquid and concave phase is gas, for example, the liquid drop in the air, the pressure of liquid is greater than the pressure of gas. If convex phase is air and concave phase is liquid, for example, the bubble in the liquid, the pressure of gas is greater than the pressure of liquid.

For cylinder surface, $r_1 \to \infty$, $r_2 = r$, then the Young–Laplace equation is changed into $\Delta p = \gamma/r$, where r is the radius of cylinder.

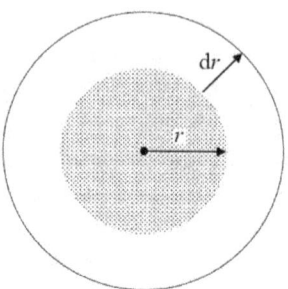

Figure 5.2. The radius for drops is r.

For sphere surface, $r_1 = r_2 = r$, then the Young–Laplace equation is changed into $\Delta p = 2\gamma/r$, where r is the radius of a sphere.

This formula can also be directly deduced from a spherical droplet. As shown in Figure 5.2, the surface area of ball with radius r is $4\pi r^2$, its volume is $(4/3)\pi r^3$, the difference of pressures which act on spherical inside and outside is supposed to be Δp, and the surface tension of the ball is γ. With the increase of pressure difference Δp, the ball expands and virtual displacement δr is produced. Then, the virtual increasing of the surface area is $\delta A = \delta(4\pi r^2) = 8\pi r \delta r$. The work for pressure difference Δp under δr is $\delta W = \Delta p \cdot 4\pi r^2 \cdot \delta r$. The new increased virtual surface free energy for overcoming surface tension is $\delta E = \gamma \cdot \delta A = \gamma \cdot 8\pi r \delta r$. Based on the conservation of work–energy, these two should be the same, $\delta W = \delta E$. Then

$$\Delta p \cdot 4\pi r^2 \cdot \delta r = \gamma \cdot 8\pi r \delta r \qquad (5.9)$$

The solution is

$$\Delta p = \frac{2\gamma}{r} \qquad (5.10)$$

For flat liquid, $r_1 = r_2 = \infty$, substituting it into the above equation leads to $\Delta p = 0$.

For plane surfaces, the pressure on the two sides of the surface is the same, i.e., there is no pressure difference. This curved surface unit is called the additional pressure of the curved liquid surface. For plane liquid surfaces, since there is no pressure difference, the additional pressure is zero. However, for convex liquid surfaces, since the pressure difference is positive, the additional pressure is also positive.

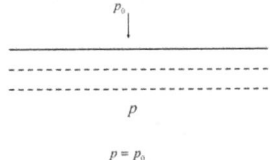

Figure 5.3. Some phenomena can be explained by the additional pressure of curved liquid surface.

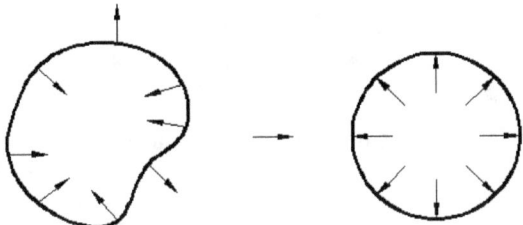

Figure 5.4. Irregular surfaces automatically shrink into a sphere.

For concave liquid surfaces, since the pressure difference is negative, the additional pressure is negative. The total pressure inside the liquid is less than the pressure in air. This condition is shown in Figure 5.3.

Free drops or bubbles present as spheres (shown in Figure 5.4). For irregular drops and bubbles, different parts of the surface have different curvature radii, and the size and orientation of the additional pressures are different. Based on Pascal's law in hydrostatics, pressures inside a drop should be the same in all respects. Hence, drops or bubbles that are acted upon by an imbalance of forces will automatically change their shapes to spheres. The additional pressures are the same, pointing inward and canceling each other out, which makes each radius of curvature the same, forming the shape of a sphere and balancing the system.

It shows superheated phenomena for liquid boiling. When the liquid saturated vapor pressure is equal to the external pressure, the temperature for gas–liquid phase equilibrium is the boiling point. The phenomenon in which a liquid exceeds the boiling point without boiling is known as superheating. During the process of boiling, vaporization simultaneously proceeds in the inside and outside of the liquid. The new formative phase is called the micro-bubble phase.

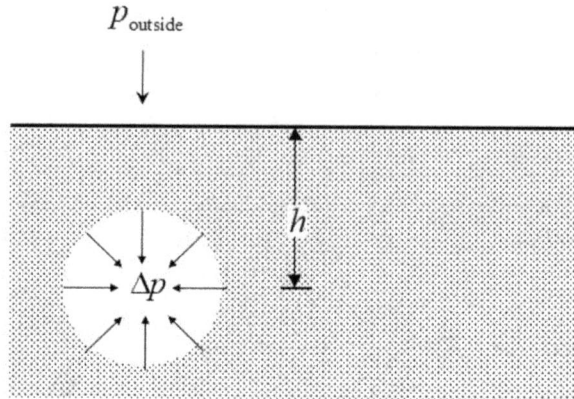

Figure 5.5.   Superheated phenomena.

Since the radius of the curvature of bubbles in liquid phase is negative, there is an additional pressure Δp which points to the inside of liquid. Thus, the bearing pressure for the bubble should be the joint force of the outside pressure, the static pressure p_h which is generated at depth h, and additional pressure, as shown in Figure 5.5.

Only when the saturated vapor pressure is equal to the joint force, bubbles become bigger and rise. Therefore, a liquid under superheating can boil only when the temperature is higher than its boiling point. If the radius of new bubbles is less than 10^{-5} m, the superheated phenomena are clear. Therefore, to prevent this, zeolites and capillaries are often added to the liquid. Air escapes from the zeolites and capillaries during the process of heating, which increases the radius of new bubbles to about 10^{-3} m so as to eliminate the superheated phenomena.

This kind of additional pressure cannot be disregarded. Two examples are as follows.

1. **Cavitation phenomenon:** At the beginning of the 20th century, when the first batch of ocean ships was manufactured successfully for a trial trip, it was found after 12 hours of trial voyage that the propellors had become tattered and could not be used. After years of research, finally, it was proved that the countless tiny water bubbles caused the damage. Usually, the damage on metal propeller by the tiny bubbles is called "cavitation".

When the propeller runs up in water, countless tiny bubbles are formed under enormous pressure. Then a sheet bubble cloud is created. Some of them are too small to be distinguished by the naked eye because of their minimal curvature radius. Experiments suggested that by now the surrounding liquid produced great pressure on the bubbles. This is an additional pressure, by which the liquid membrane of the bubbles will contract with great speed and then rupture. The pressure produced by the fracturing of the liquid membrane can be thousands of Mbar (1 Mbar $= 10^5$ MPa). Countless small bubbles with such great pressure have a continuous and intensive impact on the metal parts, which leads to damage. This is cavitation. There are many methods to avoid cavitation, for example, coating some material like sodium diethylene dithiocarbonate on the paddle surface or using the "exceeding cavitation" method, etc.

2. **Air lock:** The nurses give patients injections of various kinds of drugs. Before the injection, they must check out whether there are some tiny bubbles in the tube. If there are some small bubbles, they must be removed. This is because when the blood is mixed with little bubbles, curved liquid surfaces will be produced in blood. While slightly adding an external pressure, the curvature radiuses of the meniscus on the two sides of the bubbles are not even, as shown in Figure 5.6. The force, which will block the flow of blood, is made. Only when the additional pressure comes to a certain value, blood can begin to flow. This is the phenomenon of "air lock".

When the human body changes from a high-pressure area to a low-pressure area, the transition should be very slow because gas solubility in liquid increases according to the increase in air pressure. Therefore, in high-pressure conditions, large amounts of gas

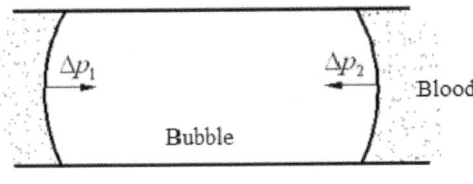

Figure 5.6. "Air lock" phenomena in blood.

dissolve in blood and tissue fluid. If the outside pressure suddenly decreases, the gas in the blood and tissue fluid will be released dramatically.

5.2   Jurin's Law in Capillary

As mentioned before, liquid rising or descending in capillary is called capillarity. If a glass capillary is inserted in water, the liquid level in the tube will rise; if it is inserted in mercury, the liquid level in the tube will descend, as shown in Figure 5.7. The surface tension of the liquid relates to the rising height h in capillary tube.

The radius of capillary tube is r and the contact angle is θ. Thus, the curved radius of liquid is $R = -(r/\cos\theta)$ and the additional pressure is $\Delta p = -(2\gamma/r)\cos\theta$.

For concave liquid surfaces, $\theta < 90°$, $\cos\theta$ is positive, and Δp is negative. In a capillary tube, the pressure below liquid surface is less than the bottom pressure because the bottom liquid has the same pressure with the outside liquid surface pressure, i.e., the atmospheric pressure. The liquid in the capillary tube is pushed upwards by the bottom pressure, causing it to rise. When the product of the rising height h and the pressure difference between the bottom of the capillary tube and the liquid surface in the capillary tube balances with

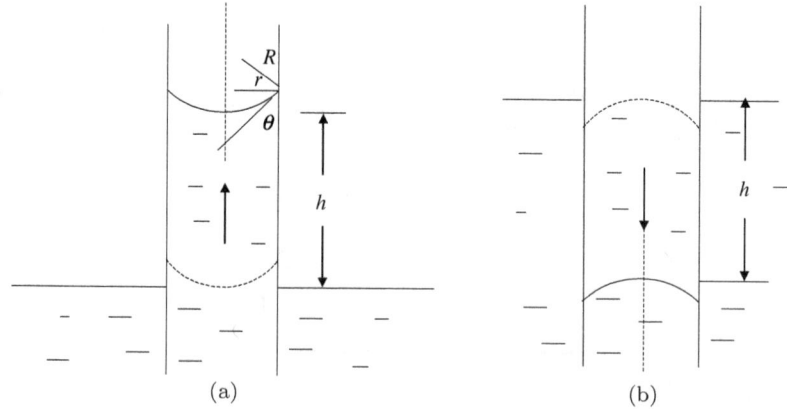

Figure 5.7.   Liquid rising (a) and descending (b) in capillary tube.

the liquid weight,

$$\frac{2\gamma \cos \theta}{r} = \rho g h \tag{5.11}$$

where g is gravity acceleration and ρ is liquid density. Then, $h = 2\gamma \cos \theta / \rho g r$. Equation (5.11) is called the capillary rising equation. After researching this kind of effect by James Jurin in 1718, it is also called Jurin's Law or Jurin's Height. It is the basic formula of measuring the surface tension by capillary tube.

For convex level, $\theta > 90°$, $\cos \theta$ is negative, Δp is positive, and the liquid surface level descends. The descending height is the same as the result from the above equation.

5.3 Liquid Bridge Force between Two Plates

It is difficult to separate two plates which have water between them, as shown in Figure 5.8. In the case of complete wetting, the contact angle $\theta = 0°$ and the side liquid surface present as concave, as shown in Figure 5.9. The pressure inside the liquid is less than the pressure of the gas outside. Since $r_1 = \delta/2$ (δ is the space between two plates), $r_2 \to \infty$, based on the Young–Laplace equation,

$$\Delta p = \gamma \left(\frac{1}{r_1} + \frac{1}{r_2} \right) = \frac{2\gamma}{\delta} \tag{5.12}$$

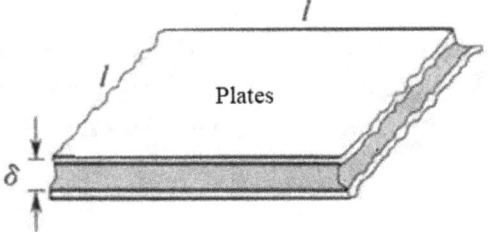

Figure 5.8. Water between two plates.

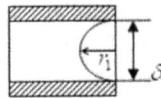

Figure 5.9. For complete wetting, side liquid surface is concave.

The force which attracts the two plates together is

$$f = \Delta p l^2 = \frac{2\gamma l^2}{8} \tag{5.13}$$

where l is the length of the plates. If the solid is irregular, it is not easy to calculate the force. However, as long as the liquid between two solids presents as concave, it is certain to have an attraction force. This kind of attraction force formed by liquid is called liquid bridge force. The existence of liquid bridges leads to a relative negative pressure and then produces the effect of attracting the two solid surfaces.

5.4   Falling Stream of Fluid Breaks up into Smaller Packets

Commonly, falling stream from faucet breaks up into droplets, no matter how smooth the stream is. This phenomenon is caused by surface tension and is also called the Plateau–Rayleigh instability.

Considering a stream of water, as shown in Figure 5.10, no matter how smooth the initial water flow is, it will always exhibit uneven thickness after being subjected to some initial disturbances. This uneven shape is manifested in two aspects geometrically. The first aspect is the radius of the water column. The second aspect is the arc shape of the outer surface of the water flow column. In fact, these two aspects deal with the two principal curvatures of the outside surface of the water column. The radius of the water column is larger in thick areas and smaller in thin areas. The outer surface of the water flow column presents an outward convex shape in thick areas and an inward concave shape in thin areas. The curvature radius of the convex shape is positive, while the curvature radius of the concave shape is negative.

According to the Young–Laplace equation, considering only the radius of the water flow column, the curvature radius in thick areas is larger, and the corresponding liquid pressure is lower. Thin areas have a smaller curvature radius corresponding to higher pressure. The high-pressure area always tries to push the liquid toward the low-pressure area. Therefore, liquids are easily squeezed from the thin areas to the thick areas, causing the thin areas to become thinner

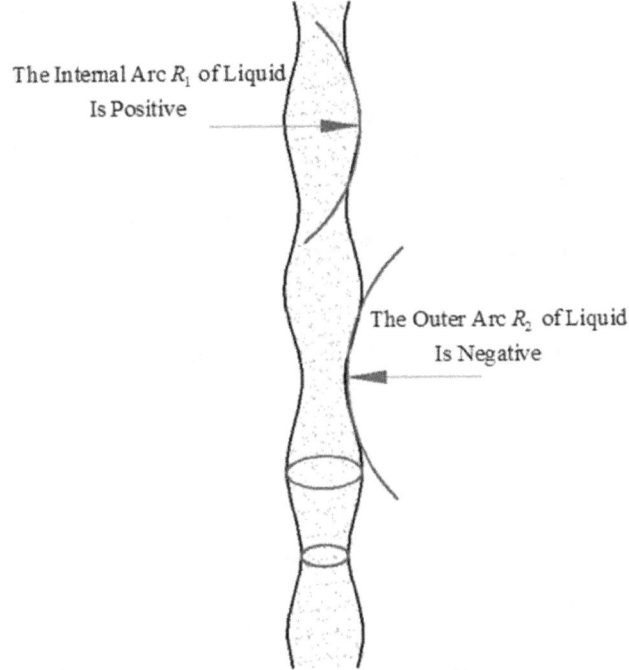

The Internal Arc R_1 of Liquid
Is Positive

The Outer Arc R_2 of Liquid
Is Negative

Figure 5.10. The uneven thickness of a stream of water after being disturbed.

and the thick areas to become thicker, resulting in neck breakage and scattering into water droplets (beads). If only this one effect exists, the phenomenon of water flow becoming water droplets is easy to understand. However, from the Young–Laplace equation, it can also be seen that, in addition to the radius of the water flow column, the curvature radius of the outer arc surface also determines the liquid pressure. From the curvature radius of the arc surface, it can be inferred that in thin areas, due to their concave shape, the liquid exhibits a relatively negative pressure to decrease the pressure, while in thick areas, due to their convex shape, the liquid exhibits a positive pressure to increase the pressure. This effect is exactly opposite to the radius effect of water flow mentioned above. These two effects usually cannot be accurately offset, and one effect is usually greater than the other. Its size depends on the initial radius and wave number of the water flow. The so-called wave number refers to the number of thick and thin peaks per centimeter of water flow, and each peak along with

a valley is recorded as a number. When the wave number is at certain
values, the curvature radius of the outer surface arc dominates, and
this effect gradually weakens over time. However, when the radius
of the water column dominates, this effect continues to strengthen.
The continuously increasing result means that the water column will
disperse into water droplets.

Although a thorough understanding of how this happens requires
a mathematical development, the diagram can provide a conceptual
understanding.

This process is an unstable process of water flow column. The
occurrence of this process is also conditional. Mathematical analysis shows that when the product of the wave number with the initial radius is less than unity ($kR_0 < 1$), the unstable components,
components that grow over time, are produced. The component that
grows in the fastest rate is the one whose wave number satisfies the
equation, $k \cdot R_0 \approx 0.697$, where k is the wave number and R_0 is the
initial radius of the water column.

By assuming that all possible components exist initially in roughly
equal (but minuscule) amplitudes, the size of the final drops can be
predicted by determining the wave number whose component grows
the fastest. As time progresses, it is the component with the maximum growth rate that will come to dominate and eventually be the
one that pinches the stream into drops.

Certainly, the Young–Laplace function can explain some other
phenomena. If a blank has been dried, the volume will be reduced.
A solid catalyst will become harder after being flooded with water.
All of this is because the wetted solids enclose a water film, and

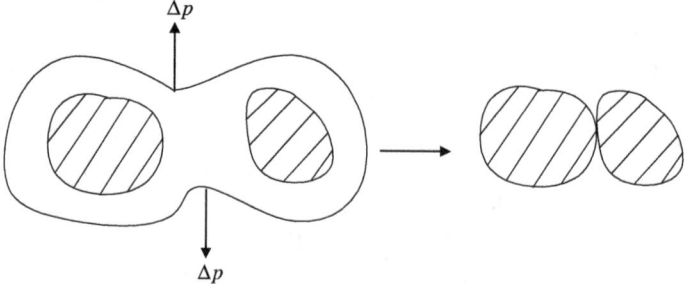

Figure 5.11.   Hardening phenomenon.

the particles are close to each other. The water film will become a slice. A concave liquid surface is formed between the particles. The additional pressure points to gas. Thus, there is a suction effect that compels the particles to integrate closely, as shown in Figure 5.11. The result is that the total volume will be reduced and the particles will cohere.

Chapter 6

Kelvin Equation

6.1 Kelvin Equation

From the Young–Laplace equation, the curvature of a liquid surface has a direct influence on inside and outside pressures. The inside pressure on a convex surface of gas phase is greater than the outside pressure. The inside pressure on a concave surface is less than the outside pressure. For a large liquid plane, since the surface is plane, the curvature is zero. Hence, there is no pressure difference between inside and outside, which means the liquid pressure is the same as the air pressure. However, for small droplets (ideal droplets are spherical, while real liquid is similar to spherical), the inside pressure is greater than the outside pressure, and the pressure difference can be expressed based on the Young–Laplace equation as $\Delta p = 2\gamma/r$, where γ is surface tension and r is the radius of liquid droplets. From this relation, the pressure of liquid droplets has an intimate connection with the radius of the droplets. The smaller the droplet, the greater the pressure difference. The bigger the droplet, the smaller the pressure difference.

The evaporation, or the process of boiling in which a liquid changes into vapor, is called gasification. Its inverse process is called condensation. In the gasification process, material molecules escape from the liquid to the vapor because of molecular heat. In the condensation process, material molecules return from the vapor to the liquid because of molecular heat. Under certain conditions (temperature, pressure, etc.), this kind of escaping and returning will reach a dynamic balance. For example, in a unit of time, the number of molecules escaping from the liquid is the same as the number of molecules returning from the vapor. Objectively, evaporation stops, and then the system is in a saturated state. The vapor is saturated vapor, and the gas pressure is saturated vapor pressure. If a certain amount of liquid is placed in an airtight vacuum container, this phenomenon can be observed. Under a certain temperature, due to molecular heat at the beginning, some of the liquid molecules can escape from the surface of the liquid. With an increase in the number of gas molecules, the density of vapor molecules increases. The vapor pressure also increases. When the temperature remains at a certain value, the gas pressure will eventually stabilize at a fixed value, and then the gas pressure is the saturated vapor pressure at this temperature. When the gas pressure reaches the saturated vapor pressure, liquid molecules are still constantly gasifying. Water molecules in the gas phase are also constantly condensing into liquid. Since the gasification velocity of water is equal to the vapor condensation velocity, the liquid does not decrease, and the gas does not increase. Liquid and gas reach equilibrium.

Put both large planar droplets and many small droplets in a closed vessel. In addition to the evaporation-saturated process described above, at certain temperatures and pressures, after a period of time, the droplets gradually disappear, and the volume of the large planar liquid increases. This phenomenon is caused by the pressure difference between the outside and inside of the liquid droplets. The large planar liquid does not have such a pressure difference. Since the outside and inside of the small spherical liquid droplets have a pressure difference and the saturated vapor is closely surrounding the droplets, the saturated vapor pressure of the small spherical droplets is actually greater than that of the large planar liquid. The difference in saturated vapor pressure leads to the disappearance of the small droplets. Since the existence of a pressure difference means the existence of a potential difference, according to the principle of minimum potential energy, if there is no other external function, the potential energy in an airtight container should tend to be consistent. Since the density of small droplets and planar liquid is the same, the potential difference mainly originates from the pressure difference. For 1 mol of liquid, its chemical potential difference can be written as $\Delta\mu^1 = \int_p^{p+\Delta p} V_m^1 dp = V_m \Delta p = 2\gamma V_m^1/r$, where V_m^1 is the mole volume of the liquid. The potential difference between the saturated vapors of small liquid droplets and planar liquid is derived not only from the vapor pressure difference but also from the change in volume. For 1 mol of vapor, its chemical potential difference can be written as

$$\Delta\mu^g = \int_{p_0}^{p_r} \frac{RT}{p} dp = V_{\mathbf{m}}\Delta p = RT\ln\left(\frac{p_r}{p_0}\right) \tag{6.1}$$

where $V_{\mathbf{m}}^g$ is the mole volume of air, R is the gas constant, T denotes temperature in degree Kelvin, p_r is the saturated vapor pressure of liquid droplets, and p_0 is the saturated vapor pressure of planar liquid.

The Kelvin function is obtained when the chemical potential energy of liquid is the same as its vapor chemical potential energy and the difference in chemical potential energy between non-planar liquid and planar liquid is the same as the difference in chemical potential energy of its corresponding saturated vapor. Under saturated conditions, since the phase change of liquid is reversible, its total Gibbs free energy is zero. Gibbs free energy can be directly used to obtain the Kelvin function. There are two ways to change the planar liquid into small droplets. First, evaporate planar liquid to saturated vapor. The saturated vapor is compressed to saturated vapor of spherical droplets, which will condense to spherical droplets. Second, the planar liquid can directly scatter into small spherical liquid droplets. Taking 1 mol of liquid as an example, the first method has three steps:

First step: Under constant temperature and pressure, 1 mol of planar liquid reversibly evaporates to saturated vapor, and p_0 is the saturated vapor pressure of planar liquid. This step is a reversible phase change. Therefore, its Gibbs free energy change is $\Delta G_{\mathrm{m},1} = 0$.

Second step: The saturated vapor of planar liquid is compressed into spherical liquid saturated vapor, which is an ideal compression process of gas under constant temperature and pressure. The Gibbs free energy change of this step is

$$\Delta G_{\mathbf{m},2} = \int_{p_0}^{p_r} V_{\mathbf{m}} dp = RT\ln\left(\frac{p_r}{p_0}\right) \tag{6.2}$$

Third step: The spherical droplets of saturated vapor condense into spherical liquid droplets. This step is a reversible phase change under constant temperature and pressure. Its Gibbs free energy change is zero: $\Delta G_{\mathbf{m,3}} = 0$.

The total Gibbs free energy change of this method is

$$\Delta G_{\mathbf{m,a}} = \Delta G_{\mathbf{m,1}} + \Delta G_{\mathbf{m,2}} + \Delta G_{\mathbf{m,3}} = \Delta G_{\mathbf{m,2}} = RT \ln \left(\frac{p_r}{p_0} \right) \tag{6.3}$$

The second method only has one step: 1 mol of planar liquid at p_0 scatters into small liquid droplets with radius r. Since there is an additional pressure, the scatter process occurs under constant temperature and inconstant pressure. The liquid pressure in liquid droplets is $(p + \Delta p)$. The additional pressure is $\Delta p = 2\gamma/r$. If we ignore the influence of liquid mole volume $V_{\mathbf{m}}^l$, the Gibbs free energy change for this method is

$$\Delta G_{\mathbf{m,b}} = \int_p^{(p+\Delta p)} V_{\mathbf{m}}^l dp = V_{\mathbf{m}}^l \Delta p = \frac{2\gamma V_{\mathbf{m}}^l}{r} \tag{6.4}$$

If the density ρ and molar mass M of the liquid are known, substituting $V_{\mathbf{m}}^l = M/\rho$ into equation (6.4) leads to

$$\Delta G_{\mathbf{m,b}} = \frac{2\gamma M}{\rho r} \tag{6.5}$$

According to the constant chemical potential energy during phase change at saturation, the chemical potential energy of the liquid droplets is equal to the chemical potential energy of the saturated vapor around them. The chemical potential energy of the planar liquid is equal to the chemical potential energy of the saturated vapor at the liquid surface. Therefore, the potential difference between the small droplets and the planar liquid should be equal to the potential difference between the saturated vapor around the droplets and the saturated vapor around the planar liquid surface, which means the two potential differences above are the same. Thus,

$$\Delta \mu^g = \Delta \mu^l \tag{6.6}$$

i.e.,

$$RT \ln \left(\frac{p_r}{p_0} \right) = \frac{2\gamma V_{\mathbf{m}}^l}{r} \tag{6.7}$$

If mole volume is substituted for mole mass M and liquid density ρ, then $V_m^l = \frac{M}{\rho}$; substitute it in the above equation:

$$RT \ln \left(\frac{p_r}{p_0} \right) = \frac{2\gamma M}{\rho r} \tag{6.8}$$

This is the Kelvin function.

The small spherical droplets show a convex liquid surface to the planar liquid. On the contrary, the liquid forms a concave meniscus surface in the pore. At this moment, the pressure inside the liquid is less than the pressure outside the liquid. Correspondingly, the saturated vapor pressure is less than that of the planar liquid.

As shown in Figure 6.1, when liquid forms a concave surface in a pore, the saturated vapor pressure p_r of the concave surface with a radius r is less than the saturated vapor pressure p_0 of the planar liquid. Liquid pressure for the concave surface is $p - 2\gamma/r$. The potential difference between the concave liquid and the planar liquid is

$$\Delta \mu^l = \mu_r - \mu = \int_p^{p-\Delta p} V_m^l dp = -\frac{2\gamma V_m^l}{r} \tag{6.9}$$

The potential difference between the corresponding saturated vapor pressure and the planar saturated vapor pressure is

$$\Delta \mu^g = \int_{p_0}^{p_r} V_m^g dp = \int_{p_0}^{p_r} \frac{RT}{p} dp = RT \ln \frac{p_r}{p_0} \tag{6.10}$$

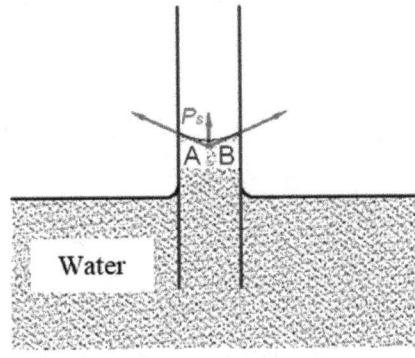

Figure 6.1. The concave meniscus formed by the liquid in the pores.

By the constant relation between the potential differences $\Delta\mu^l = \Delta\mu^g$, we have

$$RT \ln \frac{p_0}{p_r} = \frac{2\gamma M}{\rho r} \tag{6.11}$$

This is the Kelvin function for the capillary pore, which indicates that the smaller the capillary, the smaller the saturated vapor pressure balanced with that of the liquid.

Since the two methods lead to the same result, the change in the total Gibbs free energy should be the same, which is $\Delta G_{\mathbf{m,a}} = \Delta G_{\mathbf{m,b}}$. Then,

$$RT \ln \left(\frac{p_r}{p_0}\right) = \frac{2\gamma M}{\rho r} \tag{6.12}$$

The two methods are shown in Figure 6.2.

The Kelvin function shows that if the temperature is certain, saturated vapor pressure p_r of a small droplet is a function of its radius r.

For a convex surface and small liquid droplets, $r > 0$, $\ln(p_r/p_0) > 0$, which means that the saturated vapor pressure of droplets is greater than that of planar liquid under the same temperature.

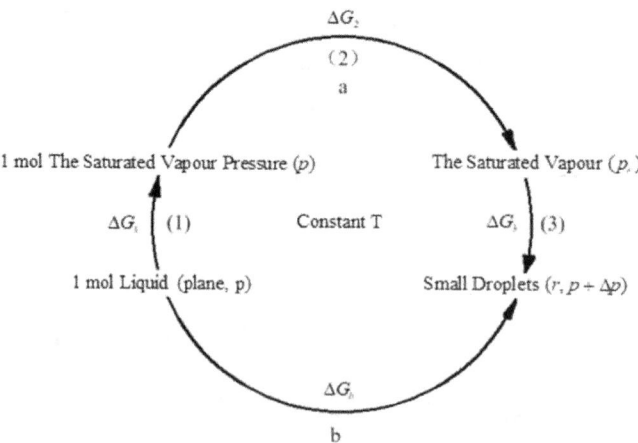

Figure 6.2. The relation between the two methods.

For a concave surface and little bubbles in water, $r < 0$. Therefore,

$$RT \ln \left(\frac{p_r}{p_0} \right) = \frac{2\gamma M}{\rho r} < 0 \qquad (6.13)$$

$\ln(p_r/p_0) < 0$, $p_r < p_0$.

Under the same temperature, the vapor pressure of liquid around the bubbles is less than the saturated vapor pressure of the planar liquid surface.

6.2 Application of the Kelvin Function

The Kelvin function can explain many phenomena and can be applied to many conditions, such as capillary pore condensation, artificial rainfall, and calculating the radius of falcate surfaces.

6.2.1 *Capillary pore condense*

From the application of the Kelvin function in concave liquid surfaces, we can find that the liquid pressure and the vapor pressure of capillary pores are less than those of a planar liquid. Therefore, under certain conditions, the gas in capillary pores is easier to condense than that outside of the capillary pores. Hence, capillary cohesion occurs when the vapor pressure is less than the saturated vapor pressure P_0, which explains why, when humidity is very high, porous media is easier to condense gas into liquid. An old saying illustrates the principle that the condensed water around a tank indicates rainy weather. Since soil and fibrous tissue have many capillary pores and water wetted in capillaries shows a concave surface, the planar liquid that has not been saturated begins to condense in the capillaries. Therefore, the capillary structure of soil can help retain water. The radius and distribution of porous solid catalysts work with the capillary condensing function, which can be calculated using the Kelvin function.

6.2.2 *Artificial rainfall*

According to the application of the Kelvin function in concave liquid surfaces, the liquid pressure and the vapor pressure of capillary pores are less than those of a planar liquid. The smaller the droplet radius (size), the bigger the pressure difference. Therefore, small droplets are easier to condense than planar liquid. From another point of view, if the gas humidity is very large, the density and pressure of saturated vapor are very large, and the saturated vapor of planar liquid is easier to condense into liquid. In the air, if the spherical liquid droplets are not spherical, which means the radius of spherical liquid droplets is zero, theoretically, the pressure difference of the inside and outside of the droplet is infinite. Therefore, it is difficult for the gas to condense to liquid directly. However, if there is an initial liquid droplet, then there is a radius of the liquid droplet, and correspondingly there is a saturated vapor pressure. When gas pressure in the air reaches the saturated vapor pressure, ambient gas will condense. With the increase in clotted liquid, the radius of the liquid droplets increases. Correspondingly, the saturated vapor pressure decreases, which implies that the condensation is easy. Therefore, in order to let the gas in the air condense, the initial droplet is the key. Without the initial droplet, no matter how much the humidity is and how large the pressure is, it is difficult to form a liquid droplet. When there is an initial droplet, under certain condition, gas will condense to liquid droplets. By utilizing this principle, when the sky has a certain meteorological condition or when the steam in clouds reaches the saturated conditions, the initial droplets can be formed by spraying micro-granules into the clouds from an airplane. It can greatly decrease the saturated degree of liquid droplets, and it becomes easy for the steam in the clouds to condense into liquid and fall to earth.

In actual cases, the steam will not condense into the liquid phase when it just reaches the saturated condition. The state of the steam needs to exceed a certain condition, which is the supersaturated state. Generally, the steam condenses into liquid in a certain supersaturated state. The saturated state is a stable state, and the supersaturation state is a kind of metastable state.

6.2.3 *Calculating meniscus radius*

According to the Kelvin function, when pressure, the saturated vapor pressure of a planar liquid, and that around the concave liquid surface are known, as well as the temperature, the mole volume of liquid, and surface tension, the following can be deduced using the Kelvin function:

$$r = \frac{2\gamma V_m^l}{RT \ln \frac{p_r}{p_0}} \qquad (6.14)$$

When there is a kind of liquid between two objects, the liquid will form a liquid bridge, and the sides of the liquid bridge will form a meniscus surface. This curved surface possesses a curvature and a curvature radius. Utilizing the Kelvin equation, the values associated with curvature or the radius of curvature can be determined.

6.3 Metastability State and the Production of a New Phase

From the Kelvin function, the saturated steam pressure of small spherical liquid or air bubbles is inversely proportional to the spherical radius. The smaller the radius, the bigger the saturated vapor pressure. In theory, when the radius is zero, the saturated vapor pressure becomes infinite, which creates the problem that it is difficult to produce a new phase. In phase-change processes, such as steam condensation, solidification of pure liquid substances, and crystallization of solute in solution, particles which produce a new phase are very tiny and the saturated vapor pressure is very large. Therefore, it is difficult to produce a new phase in the system, and supersaturation is subsequently induced.

6.3.1 *Supersaturation vapor*

The reason for the existence of supersaturated vapor is that the newly generated tiny particles of saturated vapor are greater than those of planar liquid. If the degree of supersaturation of the liquid is not high and micro liquid droplets do not reach a saturated state, the droplets cannot be generated or exist. According to the usual phase-balance conditions, the steam should condense. This steam that does not condense is called supersaturated steam. For example, near $0°C$, sometimes vapor begins to automatically condense when its pressure reaches five times the balance of vapor pressure.

When there is dust in the steam or the inside surface of the container is rough, these materials can act as sites of devaporation and make liquid droplets produce or grow easily. Even if the supersaturation degree of the liquid is very small, steam begins to condense. The principle of artificial rainfall shows that, when steam in the cloud reaches its saturation or supersaturation state, we can scatter tiny AgI particles in the cloud. Then, the AgI particles will act as sites of devaporation and cause the supersaturation degree of the new phase to decrease dramatically. Then, water in the clouds easily condenses into droplets and falls to the earth.

6.3.2   Overheating liquid

Under air pressure, when a liquid is boiling, gas not only proceeds on the surface of the liquid but also inside the liquid. If the liquid does not contain materials that can offer new phase seeds (little bubbles), even if the liquid is heated to above its boiling point, it will not boil. Therefore, based on phase equilibrium conditions, this kind of liquid is called an overheated liquid. The reason for overheating is that new phase seeds are difficult to generate in the liquid.

In order to prevent overheating, we often use dry enamelware or capillaries that contain gas because these materials can store gases. When heated, these materials release small bubbles, which help overcome the difficulty of producing new phase seeds. Therefore, it dramatically decreases the degree of overheating.

6.3.3   Subcooled liquid

Under a certain pressure, when liquid materials are cooled to their freezing point, new solid-phase particles will be generated based on the phase equilibrium condition. However, the newly generated particles are very small, and their melting point is lower. Now, small crystals do not reach their saturation state, so they cannot be produced automatically or exist. The liquid must continue to be cooled to below the normal freezing point and reach the freezing point of the small crystals. Then, the crystals will separate out. Therefore, based on the phase equilibrium condition, this kind of liquid is called a subcooled liquid. For example, pure water cannot freeze even if slowly cooled to $-40°C$. Put some small crystals in the subcooled liquid as new phase seeds; the liquid will condense quickly.

6.3.4 *Supersaturated solution*

If a solution undergoes evaporation at a constant temperature and specific pressure, the density of the solution will gradually increase. When the density reaches the saturated density of ordinary crystal solutes, according to the equilibrium condition, there should be crystal precipitation. However, since tiny crystal solutes have greater solubility, the solutes of tiny crystals do not reach their saturation state. Therefore, there is no crystal precipitation. The liquid must continue to evaporate until it reaches its saturation state. Then, there will be crystal precipitation. According to the equilibrium condition, this kind of liquid is called a supersaturated solution.

During the process of crystallization, when the liquid is evaporated to a certain saturation state, put some small crystals in the crystallization system as new phase seeds. Thus, larger crystals can be obtained.

According to thermodynamics, all supersaturated systems are not real balanced systems and in steady state. Therefore, they are called metastable states. However, these kinds of systems can continue to exist without phase changes for a long period of time. It is because under a certain condition, new phase seeds are very difficult to be generated. For example, metal quenching involves heating a kind of alloy product to a certain temperature. The temperature is maintained for a while; then, the alloy product is quickly cooled in water, oil, or other kinds of medium. Under normal temperature, the product can maintain its structure as it does in high temperatures. In this way, the performance of metal products can be improved.

Chapter 7

Surface Tension Gradient and Marangoni Effect

7.1 Surface Tension Gradient

The Young–Laplace equation reveals that for a concave liquid surface, the internal pressure within the liquid is lower than the external pressure, resulting in a state of negative pressure. This negative pressure generates a tensile force in the surrounding medium, which correlates directly with the surface tension and also directly with the curvature of the surface (inversely proportional to the radius of curvature). Consequently, as surface tension increases, so does the tensile force exerted on the surrounding medium.

For one kind of liquid, if the physical and chemical material properties (such as material constitution, density, concentration, and temperature) are uniform, it will form a single homogeneous phase. However, the stringent definition of a phase complicates the formation of a truly homogeneous phase. Distinct materials can yield different phases, while identical materials may not necessarily result in the same phase due to variations in density, concentration, or temperature. Consequently, even with identical compositions, discrepancies in these properties can lead to the emergence of multiple phases. The presence of different phases is characterized by an interface, which is associated with interfacial free energy and interfacial tension. In a liquid material, any variation in density, concentration, or temperature results in an overall non-uniformity, thereby giving rise to distinct phases. If certain physical quantities, such as density, concentration, and temperature, exhibit continuous variations within the bulk of the material, gradients will emerge — specifically, density gradients, concentration gradients, and temperature gradients. These physical gradients subsequently induce phase gradients and interfacial gradients, which can lead to gradients in surface tension or interfacial tension. Interfacial or surface tension gradients may manifest within the bulk of the material or along its surface. In a strict sense, interface tension present in the space manifests on the different interfaces in the bulk space. For example, because of different concentrations, a material with the same component will have multiple phases and interfaces. Each interface corresponds to distinct interfacial tensions, reflecting spatial changes in interfacial tension and resulting in an interfacial tension gradient. Similarly, surface tension gradients on a surface are primarily characterized by differing surface tensions across various regions of the same surface. Variations in temperature can lead to disparate temperature zones on an identical surface, which may subsequently result in changes in surface tension across the surface, thereby forming a surface tension gradient.

The phenomenon of surface tension indicates that liquids with higher surface tension exert a more substantial pulling force on their surrounding medium compared to those with lower surface tension. From a spatial perspective, the presence of a surface tension gradient implies a variation in surface tension across different regions. Specifically, areas characterized by elevated surface tension exhibit a stronger cohesive force, leading to the tendency of the liquid to migrate toward these regions while distancing itself from areas of lower surface tension. As previously mentioned, such surface tension gradients can arise from concentration gradients. For instance, in wine, the concentration of alcohol is not uniform throughout the liquid, creating both a concentration gradient and a corresponding surface tension gradient. Additionally, temperature gradients can also induce surface tension gradients. For example, if a liquid surface experiences varying temperatures, this temperature gradient will influence the surface tension, thereby resulting in a gradient in surface tension across the liquid.

7.2 Marangoni Effect

Given that a liquid with high surface tension exerts a stronger pull on the surrounding liquid compared to one with low surface tension, the existence of a gradient in surface tension will inherently induce the flow of liquid away from areas of lower surface tension. The surface tension gradient can be caused by a concentration gradient or a temperature gradient (surface tension is a function of temperature), which is called the Marangoni effect. This phenomenon was first identified in the so-called "tears of wine" by physicist James Thomson (Lord Kelvin's brother) in 1855. In a presentation at the Royal Academy of Sciences in the United Kingdom, Thomson described convection resulting from differences in surface or interfacial tension. He demonstrated this by introducing a small quantity of alcohol onto the surface of water, which resulted in a rapid surface movement from the point of alcohol introduction to the surrounding area. Subsequently, Italian physicist Carlo Marangoni investigated this phenomenon for his doctoral dissertation at the University of Pavia, publishing a related thesis in 1865 and conducting numerous studies on similar effects. Although both Thomson and Marangoni attributed the observed convection to variations in alcohol concentration affecting surface tension, the phenomenon ultimately became known as the Marangoni effect. Because most of the early treatment work performed was by Willard Gibbs, sometimes it is also referred to as the Gibbs–Marangoni effect. The reason for this effect is that liquid with high surface tension has stronger tension than the liquid around it. Therefore, the surface tension gradient will cause the liquid to flow from regions with low surface tension to regions with high surface tension. The Gibbs–Marangoni effect occurs as a result or a process. From the view of a result, it reflects one kind of effect. As a process, it reflects that liquid flows from regions with low surface tension to regions with high surface tension. Since the source of the effect is a surface tension gradient, if there is a surface tension gradient, the Marangoni flow exists. This flow can facilitate mass and heat transfer, forming a circulatory flow of a certain region, and is often referred to as Marangoni convection. Change in convection speed should occur with surface tension gradient. At the present time, mass transfer and heat transfer have extensive functions in science and technology. Therefore, much more research attention is focused on these.

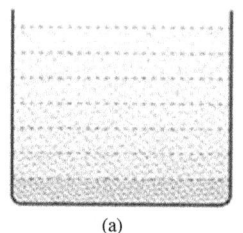
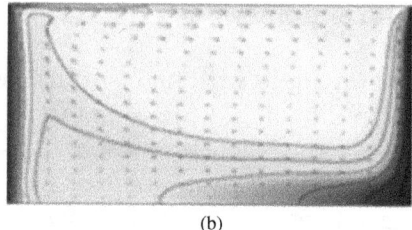

(a)   (b)

Figure 7.1. Phase boundary of the same component liquid with different concentrations and temperatures: (a) concentration gradient leads to interface tension gradient; (b) temperature gradient leads to interface tension gradient.

In the context of a liquid film, the Marangoni effect refers to the phenomenon whereby disturbances in the external environment, such as variations in temperature and concentration, lead to a reduction in thickness in certain areas of the liquid film. This alteration induces a Marangoni flow, driven by a gradient in surface tension, which facilitates the movement of the liquid toward the thinner regions along the most favorable path. Specifically, in liquid systems, the Marangoni effect indicates that the regions with high surface tension have tension against the regions with low surface tension, resulting in a liquid flow directed toward the high-surface-tension zones, as illustrated in Figure 7.1.

The Marangoni effect encompasses a variety of applications across different contexts. The most common examples are soap membrane and the tears of wine effect in a wine glass. In the case of soap bubbles, the Marangoni effect plays a crucial role in stabilizing the soap film. Similarly, in the context of wine tears, this effect facilitates the movement of high-concentration wine liquid. Furthermore, the Marangoni effect is significant in the processing of integrated circuits, particularly in the drying of silicon chip surfaces following wet-processing procedures. The presence of liquid droplets on the silicon surface can lead to oxidation, which may compromise the integrity of the material. In order to avoid forming falcate liquid bridges or droplets on the wet silicon surface, isopropyl alcohol (IPA) is dispersed as an aerial fog or stream on the wet silicon surface. This application leverages the Marangoni effect to create a gradient in surface tension, resulting in a Marangoni flow. Under the influence of gravity, this process enables the detachment of droplets from the silicon surface, ultimately leading to a dry silicon surface.

7.3 Phenomenon Explained by Marangoni Effect

7.3.1 *Soap bubble*

As shown in Figure 7.2, a soap bubble is a very thin film of soapy water which forms a sphere with an iridescent surface. Soap bubbles usually last for only a few seconds before bursting, either on their own or by coming into contact with other objects. Soap bubbles are often used to entertain children. They are also used in artistic performances. Soap bubbles can help solve complex mathematical problems of space, as they will always find the smallest surface area between points or edges.

A soap bubble is able to exist due to the surface tension of a liquid, typically water, which allows the surface layer to function similarly to an elastic membrane. Soap films exhibit significant flexibility and can generate waves in response to applied forces. However, a bubble composed solely of a single pure liquid lacks stability and requires the presence of a dissolved surfactant, such as soap, for stabilization. A common misconception is that soap increases the surface tension of water. Actually, the opposite is true. Approximately, soap decreases one-third of the surface tension of pure water. Rather than reinforcing bubbles, soap stabilizes them through the mechanism of the Marangoni effect. As the soap film is stretched, the concentration of soap at the surface diminishes, resulting in an increase in surface tension. Consequently, soap selectively fortifies the weakest regions

Figure 7.2. A soap bubble.

of the bubble, thereby preventing excessive stretching in any particular area. Additionally, soap contributes to a reduction in evaporation rates, which prolongs the lifespan of the bubbles, although this effect is relatively small.

The spherical shape of a bubble is also caused by surface tension. For a given volume, the sphere has the smallest possible surface area, which takes on a spherical shape and minimizes the free surface of a bubble.

7.3.2 *Tears of wine*

As shown in Figure 7.3, a clean and transparent glass is filled with wine. Upon remaining stationary for a duration, it was noted that the upper portion of the wine glass, which had not been in contact with the wine, retained its clarity and transparency. However, there were small droplets continuously forming and dropping back into the liquid near and above the wine level in the glass. This phenomenon is called **tears of wine**. It is most readily observed in a wine which has a high alcohol content. It is also called **wine legs**.

The observed phenomenon can be attributed to the lower surface tension of alcohol in comparison to that of water. When alcohol is homogeneously mixed with water, areas with a lower concentration of

Figure 7.3. Tears of wine.

alcohol exert a greater attractive force on the surrounding fluid than areas with a higher concentration of alcohol. The result is that the liquid tends to flow away from regions with higher alcohol concentrations. This can be easily and strikingly demonstrated by spreading a thin film of water on a smooth surface and then allowing a drop of alcohol to fall on the center of the film. The liquid will rush out of the region where the drop of alcohol fell.

Wine is composed of a combination of alcohol and water, along with dissolved sugars, acids, colorants, and flavor compounds. When the surface of the wine comes into contact with the interior of the glass, capillary action causes the liquid to ascend along the glass's surface. Both alcohol and water evaporate from this ascending film; however, alcohol evaporates at a faster rate due to its lower boiling point. This reduction in alcohol concentration results in an increase in the surface tension of the liquid, which in turn facilitates the upward movement of additional liquid from the main body of the wine, characterized by a lower surface tension owing to its higher alcohol content. As the wine ascends the glass, it forms droplets resembling tears, which subsequently fall back due to gravitational forces. This phenomenon is commonly referred to as the "tears of wine". Additionally, the elongated appearance of the wine as it climbs the glass is often described as the "legs of wine".

7.3.3 *Wafer drying in wet processing*

In the sequence of processes involved in the "wet process", drying represents the final and critical step. The drying step contributes much to the wet process performance, perhaps even defining its results. If drying is not well done, the chip cannot perform well. Therefore, the drying process is the most important process in the wetting process. A pivotal consideration is the selection of an appropriate drying method. Spin dryers and IPA are commonly used today. However, with the development of micro-technology and the requirements of increasingly high performance of chips, the size of the structure in the chip is getting smaller and smaller, and smaller particle sizes have come into focus. In terms of strength, the chip should not and cannot withstand too much mechanical stress during the processing, and residual stress is not allowed after processing.

Hence, during any process, the mechanical stress should be completely prevented from being introduced to the wafers or should be reduced to as small as possible. Therefore, the method of spin drying by a rotating machine is obviously not suitable. This more or less spells the end of spin dryers for advanced applications. Although the IPA vapor drying method is effective, it poses many safety hazards due to the need to heat the alcohol and has high requirements for fire prevention. Equipped with a large number of CO_2 fire suppression systems, the cost is also high and generally difficult to afford. Therefore, it is not very suitable for such dryness. Of course, there are also high-energy ozone drying methods, but they are not very suitable for a wide range of chip or material types. Based on analysis, the only technology that already has an inherent potential for the future is the drying technology based on the Marangoni effect. This drying technology has been well established over the past decade and accepted industry-wide. However, the commonly used technical solutions for this drying technology have also reached their limits. The low amount of IPA currently available within the process chamber and the inability to change it on a recipe base requires changes and improvements, especially for larger wafer sizes and smaller CDs. Along with the development of etching technology, crystalline chips need to be processed more carefully and dried thoroughly. Through process improvements, it is possible to further enhance the drying technology based on the Marangoni effect.

Alcohol vapor is imported into the dry chamber through atomization. The aerosol is created by an ultrasonic oscillator, which is operated on a specific resonant frequency and primarily defined by the mechanical length of a nozzle. By optimizing the minimum spray critical power, the best drying result can be obtained. Another way is to adjust the drain speed. The traditional drain speed is 0.3–2.5 mm/s. Today, the typical setting is 1–1.5 mm/s. With the availability of different drain speeds over the cross-section, different wafer surface area regions can be treated separately. The water columns on top of the wafers can be drained faster to gain time. The processes run in an ambience filled with N_2, which provides an inert atmosphere. With the very low alcohol utilization during the process, the IPA concentration stays far below its flame point, and this new technology avoids safety risks. Therefore, the adapted processes can not only avoid safety risks but also improve the drying efficiency.

7.4 Marangoni Convection

In microgravity environments, the significance of surface and interface tension is markedly pronounced. The presence of a surface tension gradient can directly induce Marangoni flow. Variations in temperature and concentration gradients can readily alter surface and interface tension. During the gas–liquid phase transfer process, the exchange of materials between these two phases results in changes to the surface tension of the liquid. When the surface tension gradient of the liquid exceeds a certain critical value, liquid regions which are close to the interface will exhibit turbulence. This interface liquid turbulence resulting from the surface tension gradient is called Marangoni convection. For a long time, it was popularly believed that Marangoni convection only exists on the interface and its intensity is small. Consequently, in analyses concerning the factors influencing actual heat and mass transfer, Marangoni convection is frequently overlooked. The Marangoni effect is often a subject of nonlinearity theory research. With further research on microscopic mechanisms of heat and mass transfer, it was found that the Marangoni effect cannot be ignored. In many heat and mass transfer processes, an abnormal phenomenon related to the transfer speed is ascribed to the Marangoni effect. In the fields of crystal production, pharmaceuticals, and metallurgy, the Marangoni effect resulting from heat and mass transfers has an important influence on the purity and quality of production. Therefore, using the Marangoni effect is a very efficient way to improve transfer effect and optimize production quality. In addition to traditional heat and mass transfer, the research scope of Marangoni convection has been extended to various fields. For example, in the process of deploying buffer solution for contact lenses, adding materials that induce the Marangoni convection effect can impel the corneal liquid to be refreshed and supplied continuously, thereby avoiding ophthalmodynia and drying the eyes. Lyford did research on potentiating of Marangoni effect to oil exploitation in petroleum of porous rock. In addition, Marangoni convection has a very important effect on many processes, such as welding and wafer drying, and also in space labs, such as heat transfer and life support systems in space, handling of material in space without containers, crystal growth, location welding, surface liquid fuel, gas desorption, and homogenization. Extensive experimentation demonstrates that Marangoni convection has an important influence on the flow state of an interface. It can promote the update of the surface liquid unit and enhance the transfer speed. The transfer speeds associated with Marangoni convection can be several times greater than those observed in the absence of this phenomenon.

Research indicates that a surface tension gradient can lead to the Marangoni effect. The formation of Marangoni convection also significantly depends on the surface tension gradient of mass transfer. The surface tension gradient determines the intensity of Marangoni convection. Since mass transfer is accompanied by remarkable interface surface tension changes, Marangoni convection caused by the transfer also exists in many mass transfer processes, and its intensity is quite considerable. When Marangoni convection occurs, according to the actual measurement by Vazquez and Lu, the interface flow speed can reach 0.1–0.6 m/s, and Marangoni convection observably enhances the transfer speed.

Currently, experimental approaches primarily focus on measuring the velocity of liquid surfaces or the macroscopic mass transfer coefficient, thereby validating the impact of Marangoni convection on macroscopic properties and mass transfer intensity. Details of Marangoni convection are difficult to get. Essentially, Marangoni convection in mass transfer is a kind of flow coupling by momentum and mass. Through fluid computations, we can avoid the difficulties of experimentation and resolution-solving and obtain simulation information about the Marangoni convection field. However, most of the simulations now depend on simplified concentration boundary conditions. Therefore, the change in surface tension with concentration in interfaces cannot be expressed, and the Marangoni convection field information cannot be extracted. Hence, until now, researchers have not been able to study the Marangoni effect completely, so the further application of the effect is limited to a certain degree. However, considering the huge practical and theoretical value of the Marangoni effect, it is highly necessary to study it further.

Marangoni convection, influenced by the induction factor acting on the surface tension gradient, can be categorized into two distinct types: temperature-driven and concentration-driven convection. The flow resulting from a concentration gradient is referred to as solute Marangoni convection, while that arising from a temperature gradient is known as heat capillary convection.

7.5 Flow State of Marangoni Convection

In order to characterize the Marangoni effect and convection, a dimensionless Marangoni number, Ma, is often used. It is found from experiments that the flow state of Marangoni convection consists of a laminar flow and an oscillation flow. What can decide the flow type is Ma. If is less than a certain critical value, $Ma < Ma_{cr}$, its flow state would remain as laminar flow. But if Ma is greater than a certain critical value, $Ma > Ma_{cr}$, the stable flow state would disappear. After a transitory stage, flow acted on by surface tension would become a kind of three-dimensional flow. The flow type can become a periodic quantity, which is related to time and space. This flow state is called oscillation flow.

7.5.1 *Flow state of laminar flow*

When a liquid or solution is introduced between two closely positioned coaxial circular disks, a liquid bridge is established. A temperature gradient can be applied within the space between the upper and lower disks, aligned with the axis of the disks. In the presence of a gravitational field, the axis of the liquid bridge should be oriented in the direction of gravity. The environment surrounding the liquid bridge may consist of liquid, gas, or vacuum. When one side of the solid structure forming the liquid bridge is subjected to heating, the temperature on that side of the liquid will increase. Given that heat transfer is a dynamic process, various locations along the two ends of the liquid bridge or the liquid surface will exhibit differing temperatures. Since temperature significantly affects surface tension, this results in variations in surface tension across different regions of the liquid surface. Consequently, a surface tension gradient is established in the spatial domain, which induces flow within the liquid. If the temperature between the two disks does not exceed a certain value and Ma is less than a certain critical value, $Ma < Ma_{cr}$, the flow would keep on being a laminar flow, as shown in Figure 7.4. The liquid close to the free surface would flow from the high-temperature region with a low surface tension to the low-temperature region with a high surface tension.

If the surface tension of one kind of liquid has a minimal value for temperature and its corresponding temperature is just right

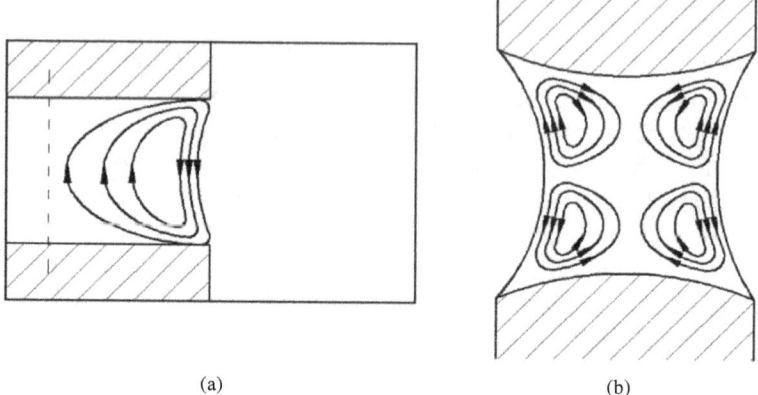

Figure 7.4. Stable flow pattern (a) top heated and (b) ring heated.

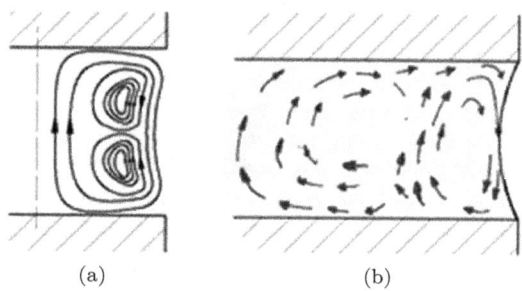

Figure 7.5. Two kinds of possible steady flow spectra.

between the top and bottom disks, such as the aqueous fatty alcohol solutions, the stream always flows from all around to the point with the lowest surface tension, not from the high-temperature regions to the low-temperature region. Chun found that the stable spectrum with a little larger Ma number would form a convective cell, which contains two cells rotating in the same direction, as shown in Figure 7.5(a).

7.5.2   *Flow state of oscillation flow*

The flow state of oscillation flow can be seen as a superposition or synthesis flow by periodic radial and circumferential motion on a vertical axis plane (Figure 7.6).

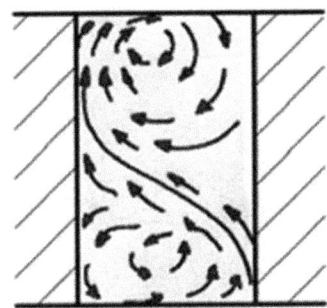

Figure 7.6. Steady flow spectrum perpendicular to the axis.

Chun and Wuest found from experiments that oscillation flow has two oscillation patterns. A decisive role is played by the scale ratio Ar. When $Ar \geq 0.5$, the flow pattern is non-symmetrical; when $Ar \leq 0.45$, it is symmetrical. Chun observed that the flow is along the circumference of the liquid bridge. Non-symmetrical oscillation flow is actually an oscillation cell. Symmetrical flow has two oscillation cells. Experimental results reported by Preisser confirmed the results from Chun, and at the same time, they pointed out that only when $n\lambda = \pi D$, oscillation exists in the liquid bridge, where λ is the circumference wavelength and n is an integer number, called modulus. A certain scale ratio corresponds with a certain modulus. A smaller scale ratio corresponds to a bigger modulus. When $n = 1$, it corresponds to the non-symmetrical flow, as referred to by Chun. $n = 2$ represents symmetrical oscillation flow. However, if the modulus is greater than 2, the flow state cannot be fixed. It is easy to change from one modulus to another. Flow patterns for these two stable modulus oscillation flows are shown in Figure 7.7. When $Ar \leq 0.5$, there exists an empirical relationship for the modulus, and the scale ratio is $nAr = 4.4$.

7.6 Research Method of Marangoni Effect

Research methods for the Marangoni effect encompass both theoretical analysis and measurement. In theoretical methods, researchers attempt to derive corresponding control functions based on the micro-mechanisms underlying Marangoni convection. Since the flow is related to nonlinear solutions, obtaining them is

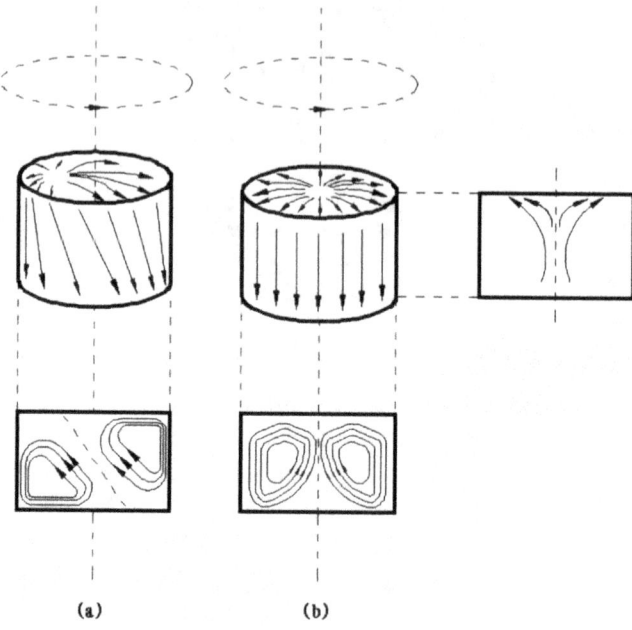

Figure 7.7. Two stable oscillation flows: (a) Module $n = 1$, nonsymmetrical oscillation flow and (b) $n = 2$, symmetrical oscillation flow.

extremely challenging. Therefore, linear methods are often employed to investigate the critical state, while nonlinear methods are used to study conditions near the critical state. With the advancement of computers and computational techniques, numerical methods have become an efficient tool for solving nonlinear problems. In experiments, there are generally two methods. One involves observing the phenomenon directly, while the other involves indirectly measuring the relevant parameters.

7.6.1   *Linear analytical method*

In general, a set of nonlinear partial differential equations, including the Navier–Stokes equation, continuity equation, and mass and heat transfer equation, can characterize the Marangoni phenomenon. Therefore, the Marangoni phenomenon is a nonlinear process that is mathematically challenging to analyze. Based on the linear stability theory of small perturbation analysis, the nonlinear term can be ignored for linearization, thereby reducing the difficulty of solving the equations. Linear stability theory serves as an efficient means for forecasting and analyzing the Marangoni phenomenon and is the foundation for subsequent nonlinear analysis. Many researchers have conducted extensive work on this basis. Sternling and Scriven were the first to theoretically investigate the impact of interface phenomena on mass transfer. They postulated that interface convection near the interface is caused by the non-uniform distribution of solute at the interface and predicted the existence of a convective structure resembling the rotary drum convection observed in heated liquid layers. Pearson conducted research on the convective conditions driven by surface tension using linear stability theory. Later, Brian considered the influence of the Gibbs adsorbed layer at the interface and extended Pearson's work. Since Brian's boundary conditions closely match real-world conditions, they are often used to study the Marangoni phenomenon driven by mass transfer.

7.6.2 *Nonlinear analytical method*

Due to the limitations of linear analytical methods, research on the Marangoni effect is confined to forecasting and analyzing the *Ma* number. Linear theory can explain convection structures, the transfer of supercritical flow types, and processes that change with time and space. Near the critical point and in supercritical regions, the nonlinear term in the governing equation has an effect that cannot be ignored. Therefore, nonlinear mathematical analysis is necessary. Since nonlinear analysis is challenging, it is often simplified as a weak nonlinear problem, focusing only on the nonlinear behavior near the critical point. At the same time, real-world applications are also substantially simplified by neglecting complex boundary conditions. Nonlinear theoretical research on Marangoni convection primarily focuses on forecasting convection structures and analysis. The results can be compared with observed structures to verify the method's validity. Scanlon and Segel were the first to investigate the nonlinear evolution of Marangoni convection in a semi-infinite atmosphere liquid–gas interface with a non-deforming interface, derived by heat transfer. Cloot and Lebon considered a more realistic and finite-depth liquid. In subsequent research, they found that the characteristic size of the convection structure is of the same order of magnitude as the liquid depth. Bestehorn studied the evolution of flow patterns in Rayleigh—Bénard—Marangoni convection, revealing the specific structure of Marangoni convection. Nonlinear analysis by Hadji showed that there are bistable regions with hexagonal or square patterns in supercritical regions. A flat interface is a common assumption for boundary conditions. However, Marangoni convection occurs at the interface, so interface deformation is a crucial characteristic that plays a significant role in the initiation and evolution of the process. Golovin considered the influence of interface deformation and predicted Marangoni convection patterns derived from temperature changes. The boundary conditions for heat transfer-induced Marangoni convection are relatively simpler than those for mass transfer-induced Marangoni convection, which does not need to account for the complex materialization of the interface adsorption layer. The application of nonlinear analysis to mass transfer-induced Marangoni convection is limited. Bragard investigated nonlinear problems related to Marangoni instability using weak nonlinear theory when a non-deforming liquid surface is absorbing liquid. They believed that a hexagonal flow structure is more stable than a rotary drum structure. Ignoring the existence of the Gibbs adsorbed layer, he studied instability problems in mass transfer-induced Marangoni convection, which includes heat effects, using the bifurcation method. By analyzing and combining nonequilibrium thermodynamics, Zhifa Sun proposed that Marangoni convection derived from mass transfer is a nonequilibrium phenomenon. It is worth emphasizing that in nonlinear theory, the wave vector plays a decisive role in selecting the flow pattern. However, since it does not account for the materialization of the liquid, it is difficult to experimentally validate the theoretical results. Approximate solutions to nonlinear problems near the critical point can be obtained. Due to the lack of real supercritical experimental data for comparison, the results cannot be easily extrapolated. At the same time, it is challenging to explain the bifurcation phenomena in supercritical regions. Although the theory of the Marangoni effect has not yet been fully developed, compared to real-world conditions, researchers can still obtain considerable useful information for further study.

7.6.3   *Numerical calculation*

With the development of modern computers and numerical simulation, it is possible to solve nonlinear numerical problems by using computer simulation. This avoids the need for analytic solutions of complicated nonlinear mathematical equations and also allows for the consideration of a number of real boundary conditions. Boyadjiev conducted research on the process of nonlinear mass transfer accompanied by Marangoni convection using the finite difference method. Zemei Tang researched liquid flow driven by the Marangoni effect on the liquid–gas surface in a liquid bridge, with the background of a floating zone, using the finite element method. Galazka studied the influence of Marangoni flow on the growth of oxidation chips. Bestehorn obtained the evolution pattern of flow structure over time, which provides direct evidence for nonequilibrium phase changes in theory. Numerical calculations can provide detailed flow information for flow and transfer processes, which remains an important way to understand Marangoni flow to this day.

7.6.4   *Displaying of Marangoni convection field*

Appealingly, the Marangoni effect can induce instability in liquid phase bulk flow and cause relative motion within the liquid. Since bulk flow disturbances and interface updates occur rapidly, it is not easy to form an ordered structure, leading to turbulence in the liquid. Therefore, it is difficult to confirm the presence of Marangoni convection. When there is no liquid bulk flow disturbance or the disturbance is minimal, the flow more often exists as a regularly ordered geometric structure. Currently, the flow conditions with the Marangoni effect can be observed by studying the Marangoni convection field. Meanwhile, by comparing the Marangoni convection geometric structures obtained from experiments with those predicted by nonlinear theoretical analytical methods, the effectiveness of these methods can be directly proven.

When Marangoni convection occurs, by adding tracers such as tiny aluminate powder, aluminum foil, or photochromic materials, the path of the tracer with Marangoni convection can be observed, and information about the Marangoni convection field can be obtained. For example, when Schwabe conducted research on

Marangoni instability in a heated liquid layer, both polygonal and radial flow structures of Marangoni convection could be observed by using aluminum foil as a tracer. Since the Marangoni effect is a phenomenon, the choice of tracer should not alter the materialization (materialization is usually used in the context of making something material or concrete, but here you might mean "manifestation" or "appearance") of the interface, so that the flow field can be obtained without disturbance.

Generally, since light has no influence on the medium, Marangoni convection at the interface can be observed using a continuous visual optical system, such as a schlieren system. Then, the Marangoni convection structure under critical instability conditions can be obtained. When using optical observation methods, the gas–liquid or liquid–liquid interface must have a visible phase boundary. This type of phase boundary is often a horizontal static surface, a horizontal surface below a strictly laminar flow, a liquid film surface with laminar flow and free from drops, or the interface of immiscible phases in droplets. These conditions can help to overcome the difficulties in observation and calculation caused by turbulent disturbances. It is easier to observe interface turbulence phenomena and more convenient to perform accurate calculations using the strict fluid mechanics functions of laminar flow.

Using the schlieren system, Orell and Westwater observed the phase interface during the liquid–liquid extraction process of glycol, acetic acid, and ethyl acetate and obtained an ordered flow structure. Imaishi observed interface turbulence phenomena when aqueous solutions of ethanolamine and diethanolamine absorbed CO_2 in a wet wall tower. They found that there were stable and continuous pine needle-like flow structures. Suciu observed the turbulent state of the phase interface in liquid–liquid extraction. Zhang conducted research on liquid–liquid extraction triggered by bi-component spreading through a liquid–liquid system. Additionally, Okhotsimskii conducted absorbing and desorbing experiments of O_2 using 16 different organic solvents in a static container and classified interface turbulence. When Agble investigated the influence of surfactants on mass transfer in a liquid–liquid system, he used the observed schlieren images as direct visual evidence of the impact of interface turbulence on mass transfer and obtained a lot of information.

Numerous experiments have clearly demonstrated that the Marangoni phenomenon exhibits a variety of flow shapes. The primary flow shapes include the rotary drum and polygon structures. While the formation mechanism of simpler flow structures, such as the rotary drum and hexagon, can be explained by weak nonlinear theory, the mechanisms of some more complicated flow structures remain unknown. In nonlinear theory, the complexity increases because the wave vector, which determines the flow shape, does not account for the material properties of the liquid. In summary, the complexity of Marangoni convection structures in different liquids reveals the nonlinear nature of the Marangoni phenomenon. More detailed experiments are needed to uncover its underlying laws.

7.6.5 *Indirect measurement method*

The appearance of the Marangoni effect in transfer processes accelerates the renewal of the liquid's surface, significantly enhancing the transfer rate. In both heat transfer and mass transfer, the prominent feature of the Marangoni phenomenon surpasses that of the mere diffusion mechanism. Consequently, by observing unusual transfer coefficients, the Marangoni phenomenon can be indirectly inferred, which not only circumvents the difficulty of directly observing the flow field but also allows for the investigation of the impact of the Marangoni phenomenon on transfer efficiency, thereby aiding in practical applications.

By measuring the macroscopic average transfer coefficient, Brian and Imaishi assessed the influence of the Marangoni phenomenon on mass transfer. The common characteristic of these measurements is to investigate the impact of interface Marangoni flow on transfer speed when the surface tension of a material changes during desorption or absorption from a solution in wet towers or similar equipment. To eliminate the influence of gravity, mass transfer is conducted in a vertical plane. Subsequently, the Marangoni effect is generated by an interface tension gradient. Lu conducted research on the impact of the Marangoni effect on liquid–gas interface transfer speeds. Tan investigated the influence of interface turbulence on liquid layer mass transfer and revised the permeate theory. Straub studied the impact on heat transfer and explained that higher heat transfer rates are associated with boiling heat transfer. Many experimental results have not only demonstrated the enhancement of the Marangoni effect on transfer processes but also quantified the critical Ma number and the relationship between transfer speed and *Ma* number by comparing measured transfer speeds with common transfer theory. Since the measured macroscopic transfer coefficient is an average of transfer coefficients and the Marangoni phenomenon is a microphenomenon occurring at the interface, information about the microcosmic transfer process cannot be directly obtained.

Laser holographic interferometer is used for microscopic quantitative measurement of the Marangoni flow phenomenon and can measure the micrometric transfer coefficient. It eliminates inaccuracies or misleading information arising from macroscopic average transfer coefficients and has considerable advantages. The Mach–Zehnder method is commonly used in interferometers. The results from this method are interference fringes. If the interference fringes are produced by the transfer process, the transfer coefficient can be obtained from the fringes. Shortcomings of laser holographic interferometers include the inability to obtain direct visual evidence, expensive equipment and consumables, and complex fringe patterns. Therefore, users require higher skills. Based on the fundamentals of interferometry, many researchers have extended this method. One classic extension is the real-time dynamic four-wavelength mixing laser interferometer proposed by Guzun-Stoica. Able suggested research combining schlieren and interference methods, utilizing the complementary advantages of both to obtain evidence of Marangoni flow and the corresponding transfer coefficient.

In addition to macroscopic parameters, such as the measurement of transfer coefficients, observation of flow structure and concentration fields around the interface, and direct observation or measurement of temperature fields, these are very important. They not only provide evidence for investigating the microscopic mechanism of the Marangoni effect but also offer practical support for understanding microscopic mass transfer and heat transfer mechanisms. Since the Marangoni effect can enhance transfer processes, discoveries from microscopic experiments are crucial for improving transfer efficiency.

As a phenomenon, the Marangoni effect and flow are directly related to interface theory and mechanical theory, and their function can be utilized in various fields. As a factor influencing liquid stability, the Marangoni effect occurring at the interface has significant impacts on chemistry, materials engineering, fluid mechanics, and space science. Current scientific development has blurred the boundaries of traditional science disciplines. Therefore, the integration of the strengths of various branches of knowledge has become a trend. As an intersection of chemistry, physics, mathematics, and mechanics, studying the Marangoni effect not only enhances our understanding of nonlinear processes but also lays the foundation for advancements in other fields.

7.7 Summary of Marangoni Convection Theoretical Analysis

In the transfer process, one of the important characteristics of the Marangoni phenomenon is that it leads to fluid mechanics instability and shapes the main flow of the liquid. Clearly, the Marangoni phenomenon has a significant influence on the transfer. In particular, it is related to the microscopic mechanism and can enhance our understanding of the transfer phenomenon. However, the microscopic instability, the nature of macroscopic flow characterization, and the evolution process cannot be fully explained through experiments alone; theories are also necessary to enable people to perfectly understand the essence of the Marangoni phenomenon.

Chapter 8

Capillary Dynamics

In previous chapters, the phenomenon of capillary rise was introduced through the analysis of statics in both theoretical and experimental contexts. However, the analysis only presents the final result of the rise, without detailing the process itself. To fully understand this phenomenon, it is essential to describe the rising process through the lens of capillary dynamics.

Many issues are associated with capillary dynamics, including the behavior of water, oil, and other liquids as they move through soil, wood, and various porous materials. This also encompasses the measurement of the density and porosity of these materials, as well as the assessment of surface tension and the viscosity of liquids.

8.1 Hagen−Poiseuille Equation

The flow of liquid in a capillary tube can be divided into several phases, among which the steady phase of liquid development is particularly important. The fluid in this phase is classified as laminar flow. Specifically, the fluid flowing through a circular tube with a uniform cross-section is referred to as Hagen−Poiseuille flow, which can be directly analyzed using the Navier–Stokes equations. The Navier–Stokes equations represent the general equations of fluid motion. Hagen−Poiseuille flow is a specific type of flow that adheres to certain assumptions, outlined as follows:

1. This flow is a steady flow and belongs to a constant flow. Therefore, for every flow parameter, its partial derivative of time is zero, $\partial/\partial t = 0$.
2. The components of flow velocity in both the radial and circumferential directions are zero; only the component along the pipeline is non-zero, $v_r = v_\theta = 0$.
3. The flow is axisymmetric and sufficiently developed. Axisymmetric means $\partial/\partial \theta = 0$, and sufficient development means that the velocity is even, $\partial v_z/\partial z = 0$.

Substituting these hypotheses into the Navier–Stokes equation with polar coordinates instead of Cartesian coordinates leads to

$$\frac{1}{r}\frac{d}{dr}\left(r\frac{dv}{dr}\right) = \frac{1}{\eta}\frac{\partial P}{\partial z} \tag{8.1}$$

When the pressure evenly changes uniformly along the circle tube, $\partial P/\partial z = \Delta p/l$, the above equation can be changed into

$$\frac{1}{r}\frac{d}{dr}\left(r\frac{dv}{dr}\right) = \frac{1}{\eta}\frac{\Delta P}{l} \tag{8.2}$$

where v represents the flow speed along the length direction of the tube (the z direction), r denotes the radial coordinate of the fluid within the tube, and l indicates the length of the fluid column in the tube. Additionally, η is the viscosity of the fluid, and ΔP is the pressure difference between the two end surfaces of l, which is generated by multiple forces.

8.2   Capillary Flow Rate

For a capillary tube, the Hagen–Poiseuille equation can be written as

$$\frac{1}{\eta}\cdot\frac{\sum P}{l} = \frac{1}{r}\frac{d}{dr}\left(r\frac{dv}{dr}\right) \tag{8.3}$$

where $\sum p$ represents the total pressure difference between the two ends of the liquid of length l in the tube. The equation above is an ordinary differential equation. After integration,

$$\frac{dv}{dr} = \frac{1}{2\eta}r\frac{\sum P}{l} + A\frac{1}{r} \tag{8.4}$$

After a second integration,

$$v(r) = \frac{1}{4\eta}r^2\frac{\sum P}{l} + A\ln(r) + B \tag{8.5}$$

A and B are unknown coefficients that can be determined using the boundary condition. According to the symmetry of the flow speed boundary and the existence of the speed grads, $A = 0$. Substituting it into expression (8.4) leads to

$$\frac{dv}{dr} = \frac{1}{2\eta}r\frac{\sum P}{l} \tag{8.6}$$

Because the radiuses of some capillary tubes are extreme small, velocity slip may occur near the capillary wall. Suppose the slip coefficient is ε. On the boundary,

$$v(R) = \frac{1}{4\eta}R^2\frac{\sum P}{l} + B = \varepsilon\left.\frac{\partial v}{\partial r}\right|_{r=R} = \varepsilon\frac{1}{2\eta}R\frac{\sum P}{l} \tag{8.7}$$

The solution is

$$B = \frac{\varepsilon}{2\eta}R\frac{\sum P}{l} - \frac{1}{4\eta}R^2\frac{\sum P}{l} \tag{8.8}$$

Substituting it into equation (8.5) leads to

$$v(r) = \frac{1}{4\eta}r^2\frac{\sum P}{l} + \frac{\varepsilon}{2\eta}R\frac{\sum P}{l} - \frac{1}{4\eta}R^2\frac{\sum P}{l} \tag{8.9}$$

where R is the radius of the circular cross-section of the tube. From expression (8.9), the flow velocity of the liquid has a relationship with the radius r. In order to describe the velocity of the whole interface of the fluid, the valid method is to use the concept of flow (volume flow) and even flow velocity. Calculate the integral of flow velocity in the section. The volume flow velocity is

$$\dot{V} = \int_0^R v \cdot 2\pi r \, dr = \int_0^R \frac{1}{4\eta} \frac{|\sum P|}{l} \left(R^2 - r^2 \right) 2\pi r \, dr$$
$$+ \int_0^R \frac{1}{2\eta} \varepsilon R \frac{|\sum P|}{l} 2\pi r \, dr \tag{8.10}$$

After integration,

$$\dot{V} = \left(\frac{1}{8\eta} \frac{|\sum P|}{l} \pi R^4 + \frac{1}{8\eta} \frac{\sum P}{l} \varepsilon \pi 4 R^3 \right) \tag{8.11}$$

Then,

$$\dot{V} = \frac{\pi}{8\eta} \frac{|\sum P|}{l} \left(R^4 + \varepsilon 4 R^3 \right) \tag{8.12}$$

The average flow velocity is the volume flow through unit area, as given by

$$\frac{dl}{dt} = \frac{\dot{V}}{\pi r^2} \tag{8.13}$$

which is also called the capillary flow rate. Substituting (8.12) into (8.13) leads to

$$\frac{dl}{dt} = \frac{\sum P}{8r^2 \eta l} \left(R^4 + 4\varepsilon R^3 \right) \tag{8.14}$$

In general, the total valid drive pressure $\sum P$ includes three types of pressures: the non-equilibrium atmosphere pressure P_A, the pressure of static water P_h, and the capillary pressure P_c. Expressions (8.12) and (8.14) are usually called the equation of capillary flow. This equation can be used to describe the liquid rate not only in a straight circular capillary tube with an even section but also in circular tubes of any shape with an even section.

The cross-sectional areas of the capillary tube, as shown in Figure 8.1, are even, and its radius is R. The capillary tube can be of any length and shape. A and B are the two ends of the capillary tube. The end A connects with the point which has the distance

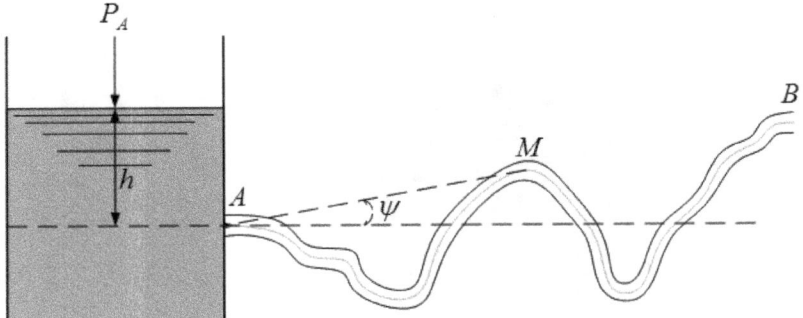

Figure 8.1. Capillary tube with arbitrary shape.

of h away from the liquid surface. The end B can be connected to the atmospheric pressure or can draw off the air from the tube and be sealed before the end A connects to the liquid. At the beginning of the flow, the resistance is very small, the acceleration is very great, and the change in flow velocity is very fast. This flow includes turbulent flow and laminar flow. After a while, the flow will change into a Poiseuille laminar flow and obey the Hagen−Poiseuille equation.

In the equation of rate, P_A can be considered a constant value; P_h can be expressed as (Figure 8.1)

$$P_{\mathbf{h}} = hg\rho - l_{\mathbf{s}}g\rho\sin\psi \tag{8.15}$$

where $l_{\mathbf{s}}$ is the straight distance from A to M, ρ is the density of the liquid, g is the acceleration due to gravity, and ψ is the angle between the straight AM and the horizontal axis. The capillary pressure P_c is

$$P_{\mathbf{c}} = \frac{2\gamma}{R}\cos\theta \tag{8.16}$$

where γ is the surface tension of the liquid and θ is the contact angle. By substituting all the above pressures into expression (8.14), the expression of the flow rate in the capillary tube can be obtained as

$$\frac{dl}{dt} = \frac{\left(P_{\mathbf{A}} + hg\rho - l_{\mathbf{s}}g\rho\sin\psi + \frac{2\gamma}{R}\cos\theta\right)}{8R^2\eta l}\left(R^4 + 4\varepsilon R^3\right) \tag{8.17}$$

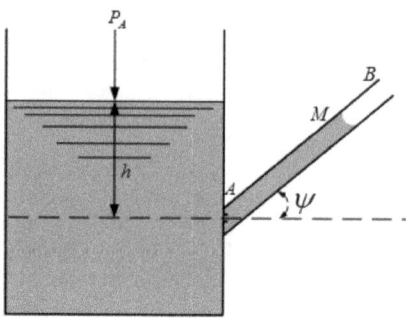

Figure 8.2. An inclined straight capillary tube.

In general, l_s, ψ, ε, and θ are functions of t. Therefore, θ may be a function of dl/dt and the pressure gradient of the liquid in the capillary tube.

Supposing ψ, ε, and θ are all constant values in expression (8.17) (note that when ψ is a constant value, the tube is not in an arbitrary shape but an inclined straight tube, as shown in Figure 8.2), in this case, $l_s = l$. Substituting them into expression (8.16) and integrating it leads to

$$\frac{\left[\left(r^2 + 4\varepsilon r\right)\rho g \sin \psi\right] \cdot t}{8\eta} + l$$

$$= \frac{P_A + \rho gh + \frac{2\gamma}{r}\cos\theta}{\rho g \sin \psi} \ln \frac{P_A + \rho g\left(h - l\sin\psi\right) + \frac{2\gamma}{r}\cos\theta}{P_A + \rho gh + \frac{2\gamma}{r}\cos\theta}$$

$$(8.18)$$

The two special cases of $\psi = 90°$ and $\psi = 0°$ are generally considered. When $\psi = 90°$, expression (8.17) can be rewritten as

$$\frac{\left[\left(r^2 + 4\varepsilon r\right)\rho g\right] \cdot t}{8\eta} + l$$

$$= -\frac{\left[P_A + \rho gh + \frac{2\gamma}{r}\cos\theta\right]}{\rho g} \ln \left[1 - \frac{\rho gl}{P_A + \rho gh + \frac{2\gamma}{r}\cos\theta}\right]$$

$$(8.19)$$

In this case, the tube is erect, as shown in Figure 8.3.

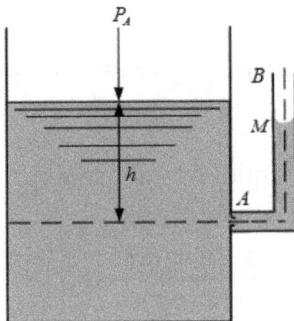

Figure 8.3. An erect capillary tube.

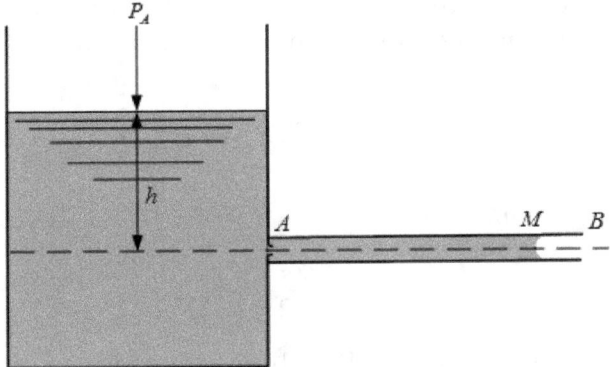

Figure 8.4. A horizontal capillary tube.

When $\psi = 0°$, expression (8.17) can be written as

$$l^2 = -\frac{\left(P_A + \rho gh + \frac{2\gamma}{r}\cos\theta\right)\left(r^2 + 4\varepsilon r\right)t}{4\eta} \tag{8.20}$$

In this case, the tube is horizontal, as shown in Figure 8.4.

If the two ends of the capillary tube are opened, $P_A = 0$, expressions (8.19) and (8.20) can be rewritten as

$$\frac{\left[\left(r^2 + 4\varepsilon r\right)\rho g\right] \cdot t}{8\eta} + l$$

$$= \frac{-\left(\rho gh + \frac{2\gamma}{r}\cos\theta\right)}{\rho g}\ln\left[1 - \frac{\rho gl}{\rho gh + \frac{2\gamma}{r}\cos\theta}\right] \tag{8.21}$$

and

$$l^2 = -\frac{\left(\rho g h + \frac{2\gamma}{r}\cos\theta\right)\left(r^2 + 4\varepsilon r\right) t}{4\eta} \tag{8.22}$$

For entirely wetting liquid, $\varepsilon = 0$, and expression (8.21) is changed into

$$t = \frac{-8\eta l}{r^2 \rho g} - \frac{8\eta\left(h + \Delta h\right)}{r^2 \rho g}\ln\left[1 - \frac{l}{h + \Delta h}\right] \tag{8.23}$$

where Δh is $2\gamma/r\rho g$.

In the case where capillary tube is entirely wetted, when ignoring the influence of pressure (e.g., $h = 0$ and the tube is horizontal) and only considering the capillary pressure, expression (8.16) can be changed into

$$\frac{dl}{dt} = \frac{r}{\eta}\frac{\gamma}{4l}\cos\theta \tag{8.24}$$

After integration, we have

$$l^2 = \left(\frac{\gamma}{\eta}\frac{\cos\theta}{2}\right) rt \tag{8.25}$$

Expression (8.25) is the Washburn equation or the Lucas–Washburn equation.

Expression (8.24) can be explained as follows. The rate of liquid penetrating the horizontal capillary tube (or a capillary tube with a small surface area) under capillary pressure is proportional to the radius of the capillary tube, the cosine of the contact angle, and the ratio between the surface tension of the liquid and its viscosity and is inversely proportional to the length of the penetrated liquid. Since $\gamma\cos\theta/2\eta$ in the Washburn equation expresses the penetration ability of liquid, it is called the penetration coefficient, or penetration efficiency. It can express the magnitude of the rate. The penetration coefficient shows the length that a liquid of unit radius penetrates the capillary tube in a unit of time. The term $\cos\theta$ in the penetration coefficient indicates that the penetration has a relationship with the material of the capillary tube.

For porous materials, the penetration rate of a liquid into the pores can be analyzed using the total volume of penetration during the period t. In order to simplify the calculation, suppose that the penetration rate of some liquid into a porous material equals the sum of n columns of capillary tubes whose radius are $r_1, r_1, \ldots, r_n$ separately.

$$V = \pi r^2 l = \frac{\pi}{2\eta^{1/2}} t^{1/2} \left(P_\mathbf{E} + \frac{2\gamma}{r} \right)^{1/2} r^3 \qquad (8.26)$$

Therefore, the total volume of the liquid penetrating the porous is

$$V = \pi \sum r^2 l = \frac{\pi}{2\eta^{1/2}} t^{1/2} \sum \left(P_\mathbf{E} + \frac{2\gamma}{r} \right)^{1/2} r^3 \qquad (8.27)$$

In this expression, the case of entirely wetting is considered. At the same time, the influence of total outside pressure is also considered. With the derivation of (8.27) with respect to time, the penetration volume in unit time, which is also called the rate of penetration volume of porous material, can be obtained.

8.3   Capillary Dynamics Equation

As shown in Figure 8.5, the liquid column undergoes three distinct types of forces during its ascent, which are capillary drive pressure $F_\mathbf{cap}$, viscous resistance on the side of the capillary tube $F_\mathbf{visco}$, and the gravity of the liquid $F_\mathbf{grav}$.

The capillary drive pressure can be directly obtained from the Young–Laplace equation. According to the Young–Laplace equation, the drive pressure in a capillary tube can be written as $\Delta P = 2\gamma/R^* = 2\gamma \cos\theta/R$. The total capillary drive pressure is $F_\mathbf{cap} = \Delta P \cdot \pi R^2 = 2\gamma \cos\theta/R \cdot \pi R^2$. Here, γ is the surface tension of the liquid, θ is the contact angle, R is the radius of the capillary tube, and R^* is the curvature radius of the meniscus.

The viscous resistance can be obtained from the Hagen–Poiseuille equation and the inner friction law for Newtonian viscous fluids. According to the inner friction law of Newton, the viscous friction shearing force on the capillary tube wall is $\tau = \eta dv/dr|_{r=R}$, where η is the viscosity of the fluid, v is the velocity of the fluid and r is the radial coordinate. It can be found that the viscous friction shearing

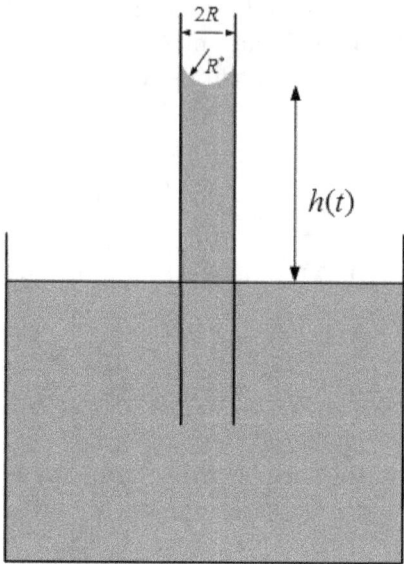

Figure 8.5.   Liquid rising in a capillary tube.

force is proportional to the viscosity and velocity gradient. The velocity gradient on the tube wall can be obtained from Hagen–Poiseuille equation. The Hagen–Poiseuille equation of a fluid flowing in a circular tube is

$$\frac{1}{r}\frac{d}{dr}\left(r\frac{dv}{dr}\right) = \frac{1}{\eta}\frac{\partial P}{\partial z} \tag{8.28}$$

where P is the pressure of the fluid and z is the coordinate in the direction of the capillary tube length, with upward considered positive. When $\partial P/\partial z$ is independent of r, integrating the above expression, for the boundary conditions $dv/dr = 0$, $r = 0$, $v = 0$, and $r = R$, leads to

$$v = \frac{1}{4\eta}(r^2 - R^2)\frac{\partial P}{\partial z} = \frac{1}{4\eta}\left|\frac{\partial P}{\partial z}\right|(r^2 - R^2) \tag{8.29}$$

$$\frac{dv}{dr} = \frac{2r}{4\eta}\cdot\frac{\partial P}{\partial z} \tag{8.30}$$

Since the flow speed is variable along the radial direction, the average flow velocity is usually used instead of the variable flow velocity.

The average flow velocity is defined as the flow per unit area, then

$$\bar{v} = \frac{\dot{V}}{\pi R^2} = \frac{1}{\pi R^2} \int_0^R 2\pi r v dr = \frac{1}{\pi R^2} \int_0^R \frac{1}{4\eta} \left| \frac{\partial P}{\partial z} \right| (R^2 - r^2) 2\pi r dr$$

$$= \frac{R^2}{8\eta} \left| \frac{\partial P}{\partial z} \right| \tag{8.31}$$

The solution is

$$\left| \frac{\partial P}{\partial z} \right| = \frac{8\eta \bar{v}}{R^2} \tag{8.32}$$

Substituting it into the velocity grads equation leads to

$$\frac{dv}{dr} = \frac{2r}{4\eta} \left| \frac{\partial P}{\partial z} \right| = \frac{2r}{4\eta} \cdot \frac{8\eta \bar{v}}{R^2} \tag{8.33}$$

At the point of $r = R$ on the parietal

$$\left. \frac{dv}{dr} \right|_{r=R} = \frac{4\bar{v}}{R} \tag{8.34}$$

Substituting this expression into the equation for the inner friction shearing force of a viscous fluid leads to

$$\tau = \eta \cdot \frac{4\bar{v}}{\gamma} \tag{8.35}$$

The total viscous resistance on the parietal of the capillary tube is

$$F_{\mathbf{visco}} = 2\pi R h \cdot \tau = 8\pi \eta h \bar{v} \tag{8.36}$$

where h is the rising height of the liquid surface. The gravity of the fluid is

$$F_{\mathbf{grav}} = \rho \pi R^2 h \cdot g \tag{8.37}$$

where ρ is the mass density of the fluid and g is the acceleration due to gravity. The resultant force of these three kinds of force is

$$F = F_{\mathbf{cap}} - F_{\mathbf{visco}} - F_{\mathbf{grav}} \tag{8.38}$$

When liquid flows in the capillary tube, not only will the velocity change (acceleration), but also the mass of the liquid will change.

Therefore, when considering the total inertial effect, these two changes must be considered at the same time. When setting up the fluid dynamics equation, the momentum law is used. According to the case of a liquid flowing in a capillary tube, its momentum law can be written as

$$\frac{d(m\bar{v})}{dt} = F \tag{8.39}$$

where m is the mass of the fluid, $m = \pi R^2 h \rho$; $\bar{v}$ is the average speed of flow, $\bar{v} = dh/dt$; F is the resultant force of capillary drive force, viscous resistance, and the gravity. Therefore,

$$\frac{d(\pi R^2 h \rho \dot{h})}{dt} = \pi R^2 \frac{2\gamma \cos \theta}{R} - 8\pi \eta h \dot{h} - \pi R^2 h \rho g \tag{8.40}$$

$$\rho \frac{d(h\dot{h})}{dt} = \frac{2\gamma \cos \theta}{R} - \frac{8\mu h}{R^2} \dot{h} - \rho g h \tag{8.41}$$

Moving the left term (called the inertial term) of the equation to (8.41) to the right side leads to

$$\frac{2\gamma \cos \theta}{R} - \frac{8\mu h}{R^2} \dot{h} - \rho g h - \rho \frac{d\left(h\dot{h}\right)}{dt} = 0 \tag{8.42}$$

According to this equation, from the aspect of the dynamics balance, capillary pressure is the driving force, but the viscous force, gravity, and inertial force are resistance forces.

8.4 Stages of Capillary Flow

With the action of capillary force, a liquid will continuously rise in a capillary tube. However, in the rising process, the liquid is influenced not only by the capillary force. In addition to it, inertial force, viscous resistance, and gravity will also act on the liquid. The rising of the liquid is the result of various kinds of forces acting together. However, at different stages, the dominant action of these forces varies. Consequently, the mechanism of liquid rising is not the same throughout the process. In the initial stage of liquid rising within a capillary tube, the volume of liquid is very small. As a result, the viscous resistance and gravitational forces, which are proportional to the volume of liquid, are also minimal and can be disregarded. However, the capillary pressure, which is related to the volume of liquid, and the inertial force, which is associated with the rate of change in the volume of liquid, cannot be ignored and will dominate. Therefore, this stage is referred to as the pure rising stage, marking the first phase of the rising process. As the liquid continues to rise, the volume of liquid in the capillary tube will gradually increase. Consequently, viscous resistance and gravitational forces will become increasingly significant, ultimately dominating the liquid's ascent over inertial forces and capillary pressure. This phase is referred to as the viscous-inertial stage, which represents the second stage of the rising process. Given that the forces in this stage are relatively greater, the dynamics of the equation will become more complex. As the liquid rises further, in the last stage of the rising process, the capillary drive pressure becomes more and more close to the viscous resistance and gravity, and the driving force and resistance become more and more close to balancing each other. Therefore, the speed of liquid rising will become slower, and the inertial effect will become weaker. The liquid flow tends to the steady Poiseuille flow. The dominant force will change from multiple forces to viscous and capillary forces. Therefore, this stage is called the pure viscous stage, which is the third stage of the whole process of liquid rising (the last stage). For various kinds of liquid rising flows, previous researchers have put forward several dynamic models. However, these models are suitable for specific backgrounds and conditions; therefore, none of them can be adapted to the whole rising process. Each model is only adapted to one stage. Since the dominant forces mostly exists in the second stage, and the related mathematical expression is the most complex one, it is divided into three stages. The temporal delineation of the three stages can be determined based on the corresponding discriminant criteria.

8.4.1 *Phases of capillary flow*

8.4.1.1 *Stage of inertial force action*

At the instant of the capillary tube coming into contact with the liquid, that is, the initial stage, the amount of liquid is very small, the viscous resistance and gravity are also very small, which can be ignored, and the dominating forces are inertial force and capillary pressure in this stage. In this case, Quere put forward a model ignoring the viscous resistance and gravity. Neglecting the viscous resistance and gravity in expression (8.42) leads to

$$\frac{2\gamma\cos\theta}{\rho R} = \frac{d(h\dot{h})}{dt} = \dot{h}^2 + h\ddot{h} \qquad (8.43)$$

By solving this nonlinear differential equation, Quere gave the rising height of capillary under a constant speed as

$$h = t\sqrt{\frac{2\gamma\cos\theta}{\rho R}} \qquad (8.44)$$

This equation is adapted to the first stage of liquid rising, the pure inertial stage.

8.4.1.2   *Stage of inertial-viscous force*

As the liquid continuously rises, the amount of liquid in the capillary tube will gradually become greater. Therefore, the viscous resistance and gravity will become more and more important, which cannot be ignored. They will dominate the liquid rising with the inertial force and capillary drive pressure. Considering that the gravity is still relatively small at the beginning of this stage, which can also be ignored, Bosanquet put forward a model ignoring the gravity. By neglecting gravity, expression (8.42) leads to

$$\frac{d(h\dot{h})}{dt} + ah\dot{h} = b \qquad (8.45)$$

where $a = 8\eta/R^2\rho$ and $b = 2\gamma\cos\theta/R\rho$.
    The solution is

$$h^2 = \frac{2b}{a}\left[t - \frac{1}{a}(1 - e^{-at})\right] \qquad (8.46)$$

This equation is adapted to the cases where the liquid surface is not so high.

8.4.1.3 *Stage of viscous force*

With viscous resistance increasing, it increasingly offsets the capillary drive pressure and the net driving force becomes smaller and smaller. Therefore, the inertia of liquid flow becomes smaller and smaller. For this case, Lucas and Washburn put forward a model ignoring the inertial effect and gravity. Neglecting the influence of inertial effect and gravity in expression (8.42) leads to

$$h^2 = \frac{\gamma \cdot R \cos \theta}{2\mu} t \qquad (8.47)$$

8.4.1.4 *Stage of viscous force and gravity*

However, when the height of the liquid surface is greater than a specific value, the influence of gravity cannot be ignored. Fries and Dreyer analyzed and demonstrated that the influence of gravity and viscous forces must be considered when $h > 0.1h_{eq}$, where h_{eq} is the height when the capillary pressure is the same as the static water pressure. Washburn gave the invisible solution for the height of capillary rising when considering the influence of viscous force and gravity, which is

$$t(h) = -\frac{h}{\beta} - \frac{\alpha}{\beta^2} \ln \left(1 - \frac{\beta h}{\alpha} \right) \qquad (8.48)$$

Fries and Dreyer gave the visible solution, which is

$$h(t) = \frac{\alpha}{\beta} [1 + W(-e^{-1 - \beta^2 t/\alpha})] \qquad (8.49)$$

where $\alpha = \gamma R \cos \theta / 4\eta$ and $\beta = \rho g R^2 / 8\eta$. $W(\alpha)$ is the Lambert W function, and it is the $W(z)$ in the equation $z = W(z)e^{W(z)}$.

When the capillary pressure is the same as the static water pressure,

$$h_{eq} = \frac{\alpha}{\beta} = \frac{2\gamma \cos \theta}{\rho g R} \qquad (8.50)$$

It is regrettable that the solution for the height of liquid rising — whether in the form of an invisible or visible solution — is derived by neglecting inertial forces due to mathematical complexities. To date, no researcher has provided an analytical solution that simultaneously accounts for inertial forces, viscous forces, gravity, and capillary pressure.

8.4.2   *Division of the different stages*

As mentioned above, the role of various kinds of forces in the total effective drive pressure has primary and secondary points in different stages of capillary flow. The process of capillary flow can be divided into three major stages in terms of the role level of these forces:

1. stage of inertial force action,
2. stage of inertial force and viscous force action together, and
3. stage of viscous force action.

The second major stage can be divided into three stages, and the whole process can be divided into five stages. It is the inertial stage in the time range of $0–t_1$ (the first major stage), where t_1 is the time boundary point of the inertial stage and the viscous stage, which can be determined using the rule that the height difference of the liquid surface rising between the inertial phase and the viscous-inertial phase should be smaller than 3%. $t_1 - t_{2,S}$ is the second stage and $t_{2,S}$ is the boundary point of the second stage and the third stage, which can be determined using the rule that the flow velocities calculated by the Quere equation and the Lucas–Washburn equation separately should be equal. $t_{2,S} - t_{2,Q}$ is the third stage and $t_{2,Q}$ is the boundary point of the third phase and the fourth stage, which can be determined using the rule that the height of liquid surface in the Quere equation and the Lucas–Washburn equation should be equal. $t_{2,Q} - t_3$ is the fourth stage and t_3 is the boundary point of the fourth stage and the viscous force stage, which can be determined using the rule that the rising height of the liquid surface calculated separately using the Bosanquet model and the Lucas–Washburn model should be smaller than 3%. $t_3 - t_{\rightarrow\infty}$ is the stage of viscous force action. For t_1, in terms of the rule mentioned above, it is

$$0.03 = \frac{h_{\mathbf{Quere}}(t_1) - h_{\mathbf{Bosanquet}}(t_1)}{h_{\mathbf{Quere}}(t_1)} \tag{8.51}$$

Tidying up, it leads to

$$t_1 = \frac{0.1856}{a} = \frac{0.0232R^2\rho}{\eta} \tag{8.52}$$

From equation (8.46), we have

$$h_1 = \frac{0.18\sqrt{b}}{a} = 0.0318\sqrt{\frac{R^3 \rho \gamma \cos\theta}{\eta^2}} \tag{8.53}$$

For $t_{2,S}$, when the flow speeds separately in the Quere equation and the Lucas−Washburn are equal, the boundary point of time is solved as

$$t_{2,S} = \frac{1}{2a} = \frac{R^2 \rho}{16\eta} \tag{8.54}$$

For $t_{2,Q}$, when the heights of liquid surface separately in the Quere equation and the Lucas−Washburn equation are equal, as shown in Figure 8.6,
the time point is solved as

$$t_{2,Q} = \frac{2}{a} = \frac{R^2 \rho}{4\eta} \tag{8.55}$$

The simultaneous equations of (8.44) and (8.47) leads to

$$h_{2,Q} = \frac{2\sqrt{b}}{a} = 0.3536\sqrt{\frac{R^3 \rho \gamma \cos\theta}{\eta^2}} \tag{8.56}$$

For t_3, when the height difference of the liquid surface rising calculated separately using the Bosanquet model and the

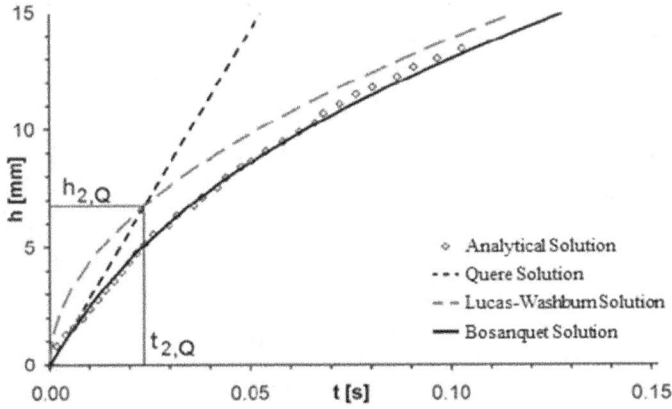

Figure 8.6. Time step for several equations.

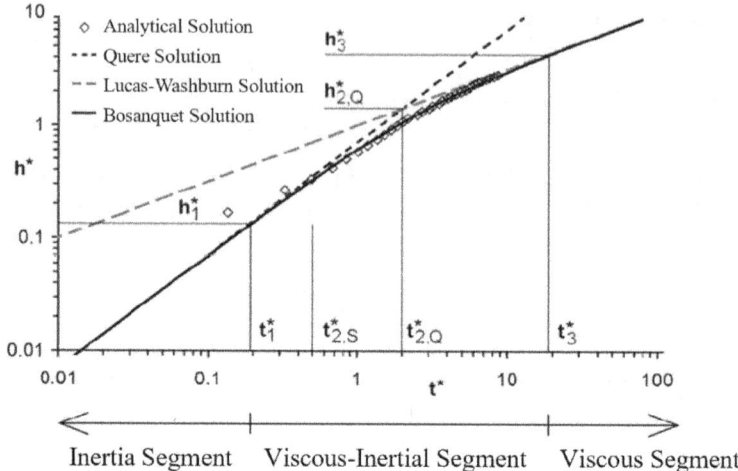

Figure 8.7.   Time point without dimension.

Lucas−Washburn model is smaller than 3%. We have

$$0.03 = \frac{h_{\textbf{Lucas−Washburn}}(t_3) - h_{\textbf{Bosanquet}}(t_3)}{h_{\textbf{Lucas−Washburn}}(t_3)} \tag{8.57}$$

then

$$t_3 = \frac{16.921}{a} = \frac{2.1151 R^2 \rho}{\eta} \tag{8.58}$$

Substituting it into equation (8.46) leads to

$$h_3 = \frac{5.6429\sqrt{b}}{a} = 0.9975\sqrt{\frac{R^3 \rho \gamma \cos\theta}{\eta^2}} \tag{8.59}$$

Ichikawa and Satoda gave the boundary point of time without dimension, as shown in Figure 8.7:

$$t^* = at = \frac{8\eta t}{\rho R^2} \tag{8.60}$$

$$h^* = \frac{ah}{\sqrt{2b}} = \sqrt{\frac{16\eta^2 h^2}{\rho R^3 \gamma \cos\theta}} \tag{8.61}$$

where $a = 8\eta/R^2\rho$ and $b = 2\gamma\cos\theta/R\rho$.

Chapter 9

Concept and Function of Liquid Bridge

9.1 Concept of Liquid Bridge

When a small amount of liquid exists between the surfaces of two closely spaced solids that can be wetted by the liquid, a connected system is formed among the two solids and the liquid. This system is called a liquid bridge. Liquid bridges have two main features: transmission and connection. Transmission refers to the convection of mass and heat transfer within the liquid. Connection refers to the pulling force exerted by the liquid on the surfaces of the two solids. Both the convection of mass and heat transfer within the liquid and the pulling force on the solids are based on the surface tension acting on the sides of the liquid.

9.2 Formation of Liquid Bridge

There are two approaches to forming a liquid bridge: one is natural, and the other is artificial. Naturally formed liquid bridges are common in micro- and nanoscale systems. Artificial liquid bridges are usually created for research on Marangoni convection during mass and heat transfer processes.

As a new mechanical and electrical system, Micro/Nano-Electro-Mechanical Systems (M/NEMS) are widely used in both civilian and military fields and have enormous potential. However, due to

181

the particularities of their size and processing, it is easy for liquid bridges to form, either during processing or in the application process. In general, liquid bridges are harmful to the performance of M/NEMS.

The processing of MEMS structures includes two types: wetting processing and drying processing. However, with the development of microelectronics and MEMS technology, the wafer size has increased significantly, from 100 to 300 mm. At the same time, the characteristic scale of process nodes has become smaller and smaller, reaching as low as 90 nm, sometimes even 65 nm or smaller. Additionally, the process depth has increased, with a standard depth-to-width ratio of 50:1 or 60:1, sometimes even 100:1. Consequently, the wafers processed by semiconductors are becoming larger and larger, with nodes becoming smaller and requiring greater precision. The structures are becoming more complex, and fewer and fewer processing methods can fulfill all these requirements. At present, the most useful method is still wetting processing, which is expected to remain the leading method in the next decade. The processing of wafers includes Front End Of Line (FEOL) and Back End Of Line (BEOL). FEOL is mostly concerned with the structure of the device, while BEOL focuses on the connection of the device structure. The surfaces that need to be cleaned in FEOL are Si or SiO_2, for which many methods can be used. However, for BEOL, fewer methods can be used due to the metal layer on the wafer.

Solid surfaces can be divided into hydrophilic and hydrophobic types. The surface of SiO_2 is hydrophilic, making it easy to wet with a cleaning agent during the cleaning process. However, the surface of Si is hydrophobic, which makes it difficult to clean because the cleaning agent cannot wet the surface and remove particles. During processing, integrated circuits (ICs) produce dirt, which can be classified into five types: general particles, metal shavings, organic matter, natural oxidation matter, and micro-roughness. These types of dirt must be cleaned off after wet processing. The cleaned wafer must meet certain standards. To achieve these standards, a series of chemicals are introduced during the cleanup process. The cleanup methods include RCA cleanup, IMEC cleanup, and Ohmi cleanup. Either before or after cleanup, some liquid (e.g., deionized water) may remain on the wafer surface and within the processed structure.

The liquid residue on the surface affects wetting and adhesion. The residual liquid within the structure will form a liquid bridge. A mild liquid bridge can affect the performance of MEMS, while a severe one can cause system failure due to adhesion.

Besides forming liquid bridges during the machining process, MEMS devices are also prone to forming liquid bridges during packaging, storage, and operation. Packaging is a crucial step for MEMS, which involves forming the tube body, connecting the pins (interface between macro and micro scales), and sealing the chip. Sealing can be either vacuum sealing or non-vacuum sealing. Currently, MEMS devices are among the most specialized for packaging. Since inert gases can alter the mechanical movement of devices during high-speed vibration, certain types of devices must operate in a vacuum to exhibit specific functions. This is not a new issue, as many electronic and optoelectronic devices and systems also require a vacuum environment. However, MEMS devices differ from traditional electronic and optoelectronic devices. Besides functioning as conductive circuits, MEMS devices can also perform mechanical movements. Due to this movement function, they require not only reduced humidity and damping but also reduced wear and increased lubrication. To a certain extent, these two requirements are contradictory, posing a new challenge for packaging techniques. Practices with MEMS devices have shown that absolute vacuum packaging is not the ideal solution. In fact, maintaining an absolute vacuum package over a long period is not only difficult but also unnecessary. Therefore, non-vacuum packaging is commonly used. Liquid bridges cannot be avoided entirely. The structures of MEMS devices are very small, and most of them contain grooves, which act as natural capillary tubes. When the air humidity is relatively high, moisture in the grooves easily condenses into droplets, forming liquid bridges.

In order to avoid residual liquid during the process and the formation of liquid bridges, the final step of the wetting process must involve removing humidity or drying. Otherwise, the processed wafer will not meet the required standards. To a certain degree, the drying process is the crucial final step that determines the overall performance of the entire wafer. If it is not dried properly, the wafer's properties will suffer. Therefore, the drying process is the most important step in the wetting process. In order to prevent the formation of droplets and liquid bridges after packaging, a relative vacuum package is necessary. However, since it is difficult to maintain a vacuum package for the device over a long period, it is essential to keep the storage environment dry and sealed. Sometimes, drying and sealing need to be processed simultaneously.

A man-made method of building a liquid bridge typically involves using two coaxial round plates that are very close to each other. Some liquid or solution is injected between these two plates. Usually, a temperature field is applied along the axial direction of the upper and lower plates. If in a gravity field, the axis of the liquid bridge aligns with the direction of gravity. The environment surrounding the liquid bridge can be either liquid, gas, or vacuum. When one side of the liquid bridge, which is in contact with a solid, is heated, the temperature of one end of the liquid will rise. This is a heat transfer process in which the temperature is inconsistent at both ends and even at different points on the liquid surface. Since temperature has a significant effect on surface tension, the latter varies in different positions or fields. Therefore, there is a surface tension gradient in space. According to the Marangoni effect, liquid tends to flow from a low-surface-tension field to a high-surface-tension field, which results in Marangoni convection. Since surface tension decreases as the temperature rises, the liquid will flow from the high-temperature field to the low-temperature field. In the liquid-bridge system, the liquid on the surface flows from one end to the other, forming convection. The man-made liquid bridge can be used for research into convection. Therefore, this type of liquid bridge is beneficial.

However, the natural formation of liquid bridges in MEMS is usually harmful. Depending on the performance requirements, when two solid components undergo relative movement or vibration, such as the relative vibration between a cantilever beam and its substrate, the liquid present between them generates an additional adhesion force. When this adhesion force is relatively small, it directly affects the amplitude of the vibration. If the adhesion force is relatively large, it can cause these two solid components to lock together and result in adhesion. Consequently, the structure becomes immobilized, and the system fails. Since many structures in micro/nano systems rely on relative movement, the presence of liquid bridges directly impacts the system's performance.

9.3 Pull-Off Force Resulted from the Liquid-Bridge

9.3.1 *Concept of pull-off force*

The force that pulls two solids together in a liquid-bridge system, exerted by the liquid, is called the pull-off force. In fact, it is also the force that needs to be overcome when separating the surfaces of two solids. The reason for the pull-off force is the surface tension on the curved liquid surface, which stems from the adhesion between the liquid and the solid, as well as the cohesion of the liquid. When liquid comes into contact with a solid that can be wetted, there will be adhesion between the liquid and the solid surface. Additionally, the liquid has cohesion, which causes a pulling effect on the two surfaces of the solids in a liquid-bridge system. The strength of the pull is determined by both the adhesion force and the cohesion force, but the maximum force (peak) is limited by the weaker of the two forces. Experiments show that when the liquid is water and the solid has a hydrophilic surface, the adhesion force is stronger than the cohesion force. In this scenario, if the two solid surfaces are moved to pull the liquid, the water is pulled off first, rather than the interface between the water and the solid. When the liquid pulls the two solid surfaces, the force is directed along the normal to the solid surface. For parallel plates, the direction of the force at every point in the affected zone is the same, and the direction of the resultant force is consistent with the direction at every point. However, for non-planar solid surfaces (such as spherical surfaces), the normal direction of different points in the affected zone differs, and the direction of the resultant force depends on the vector sum of the forces at those points. The pull-off force is the force that needs to be overcome to separate the two solids due to the adhesive force. For the convenience of analysis, the line is defined as the axial line of the liquid bridge, which passes through the liquid and is perpendicular to the two solid surfaces. The adhesive force, which acts along the axial line of the liquid bridge and is exerted by the liquid on the solid surface, contributes to the pull-off effect.

9.3.2   *Formation mechanism of pull-off force*

In a liquid-bridge system, the adhesive effect exists between the liquid and the solid, and the liquid exhibits a cohesive effect. Therefore, the liquid pulls the two solid surfaces together, which is the essential reason for the occurrence of the liquid bridge. The pull force exists at any scale because adhesion and cohesion exist at any scale. However, the pull-force effect is not always obvious at every scale. If the span (the distance between the two solid surfaces) of the liquid bridge is relatively large or the overall size of the liquid-bridge system is relatively large, the effect of the pull-off force is not obvious. The reason for this is that other forces (such as volume forces) will dominate. Such an effect is similar to the capillary effect; therefore, the theory of the capillary effect is often used to analyze the pull-off force. In fact, the characterization mechanism of the liquid bridge can be explained by the effect of the capillary force. Analyzing the capillary effect, the two ends of the liquid connect with the solid surface, exhibiting an adhesive effect. There is surface tension on the side surface of the liquid, and its combined effect forms a meniscus and an inside concave surface of the liquid. This surface causes a pressure difference between the two sides of the liquid surface (one side is the gas outside the liquid, and the other side is the liquid itself). Due to the concave surface, the pressure outside the liquid (gas) is higher than that inside the liquid, creating a negative pressure within the liquid. This negative pressure results in a pumping force that pulls and attracts the surrounding medium (the upper and lower solids and the gas on the side surface). The component of this pumping force that is vertical to the solid surface is the pull-off force.

9.3.3   *The analysis model of pull-off force*

There are two approaches to analyzing and determining the pull-off force model: one is the thermodynamic method, and the other is the molecular theory method. The thermodynamic method encompasses two models: one is the capillary theory model, which is based on the capillary pressure difference equation derived from the Young–Laplace equation; the other is the energy theory model, which is grounded in the principle of conservation of work and energy.

9.3.3.1 *Capillary theory model for liquid bridge*

1. Liquid bridge between two parallel plates

As shown in Figure 9.1, when a small amount of liquid exists in a narrow gap between two parallel solid plates, a liquid-bridge system is formed. Since adhesive forces exist between the liquid and the two solid surfaces, and the liquid possesses cohesive force, the two parallel solid plates are adhered together. Consequently, an interaction and interconnection system is established. Due to the adhesive effect of the solids on the two ends of the liquid's side surface, combined with the liquid's surface tension, a meniscus is formed. This shape results in a pressure difference between the interior and exterior of the liquid, with the pressure of the gas outside the liquid being higher than that inside, thereby creating a negative pressure within the liquid. According to the capillary theory model, the pull-off force has two components: one is the effect of the negative pressure inside the liquid acting on the wetted area of the solid in the direction of the liquid-bridge axes; the other is the action of the surface tension of the liquid's side surface along the liquid-bridge axes at the three-phase lines.

Assume that A_t and A_b represent the wetted area of the upper and lower solid surfaces. The distance between the two plates is d. The curvature radius of the liquid surface is r. Then, according to the Young–Laplace equation, the pressure difference between the two sides of the liquid surface is

$$\Delta p = \gamma \left(\frac{1}{r_1} + \frac{1}{r_2} \right) \tag{9.1}$$

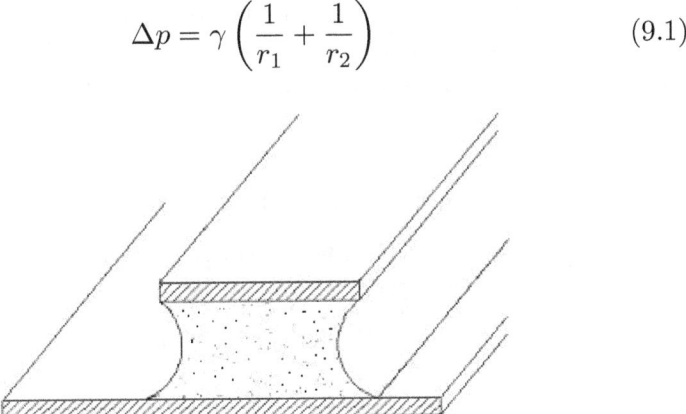

Figure 9.1. Diagram of liquid-bridge.

where γ is the surface tension of the liquid, r_1 is the principal curvature radius of the side curve surface of the liquid on the plane which is vertical to the plate, and r_2 is its principal curvature radius on the plane which is parallel to the plate. For the square plate, the wetted area is also square and the boundary is a line. Therefore, $r_2 = 0$. While r_1 has a relation with d, the distance between the two plates. Assume that the contact angles of the liquid with the upper and lower plates are, respectively, θ_t and θ_b. According to the geometry relation, $d = r_1 \cos\theta_t + r_1 \cos\theta_b$. Substituting it into equation (9.1) leads to

$$\Delta p = \frac{\gamma(\cos\theta_t + \cos\theta_b)}{d} \tag{9.2}$$

According to the Pascal principle, this negative pressure acts not only on the side surface of the liquid but also on the wetted surface of the upper and lower solid surfaces. The total forces act on the wetted areas of the upper and lower solid surfaces are, respectively,

$$f_{t1} = \Delta p A_t = \frac{\gamma A_t(\cos\theta_t + \cos\theta_b)}{d} \tag{9.3}$$

$$f_{b1} = \Delta p A_b = \frac{\gamma A_b(\cos\theta_t + \cos\theta_b)}{d} \tag{9.4}$$

The other part of the pull-off force is the component of surface tension of the liquid side surface acting on the three-phase lines. It is vertical to the parallel-plate surface. If l_t and l_b represent the circumference of the wetted area of the upper and lower solid surfaces, respectively, the total force acting on the boundary of the wetted areas of the upper and lower plates are, respectively,

$$f_{t2} = \gamma l_t \sin\theta_t \tag{9.5}$$

$$f_{b2} = \gamma l_b \sin\theta_b \tag{9.6}$$

Adding these two parts, the total pull-off force of the upper and lower plates are, respectively,

$$f_t = f_{t1} + f_{t2} = \frac{\gamma A_t(\cos\theta_t + \cos\theta_b)}{d} + \gamma l_t \sin\theta_t \tag{9.7}$$

$$f_b = f_{b1} + f_{b2} = \frac{\gamma A_b(\cos\theta_t + \cos\theta_b)}{d} + \gamma l_b \sin\theta_b \tag{9.8}$$

These are the force and the counterforce. Therefore, $f_t = f_b$. When the material and size of the upper and lower plates are the same, then $\theta_t = \theta_b = \theta$, $A_t = A_b = A$, $l_t = l_b = l$, and

$$f = \frac{2\gamma A \cos\theta}{d} + \gamma l \sin\theta \qquad (9.9)$$

For non-square plates (such as circular plates), the wetted area is not a square and the boundary line is not a straight line. Therefore, $r_2 \neq 0$. In this case, the effect of r_2 should be taken into account. The non-square surface of the solid can be analyzed similarly as above. However, in general, it is more complex.

2. Liquid bridge of spherical plates

In the liquid-bridge system, the upper surface of the solid can also be spherical. Assume the radius of the spherical surface is R. The two principal curvature radiuses of the liquid meniscus are r and r^*. The contact angle of liquid with the upper surface is θ_t. The contact angle of liquid with the lower surface is θ_b. The surface tension of liquid is γ. The angle at the position where the liquid comes into contact with the wetted three-phase lines of the spherical surface is ϕ. The distance between the bottom end of the sphere and the bottom plane is d. Then, according to the Young–Laplace equation, the pressure difference between the inside and outside of the meniscus side surface of liquid is

$$\Delta p = \gamma \left(\frac{1}{r} + \frac{1}{r^*} \right) \qquad (9.10)$$

When the size of the sphere is relatively larger than the liquid (when $R \gg r$, $r^* \gg r$), the effect of principal curvature radius can be ignored. The above equation can be written as

$$\Delta p = \frac{\gamma}{r} \qquad (9.11)$$

According to the geometric relation, it is known that

$$R(1 - \cos\phi) + d = r[\cos\theta_b + \cos(\theta_t + \phi)] \qquad (9.12)$$

Then,

$$\frac{1}{r} = \frac{\cos\theta_b + \cos(\theta_t + \phi)}{R(1 - \cos\phi) + d} \qquad (9.13)$$

Substituting it into the pressure difference equation leads to

$$\Delta p = \frac{\gamma[\cos\theta_b + \cos(\theta_t + \phi)]}{R(1 - \cos\phi) + d} \qquad (9.14)$$

By multiplying the negative pressure by the projected area of the wetted zone of the spherical surface onto a plane vertical to the bottom plane, the pull-force along the axes of the liquid bridge is determined to be

$$f_1 = \Delta p \pi R^2 \sin^2\phi = \frac{\pi R^2 \sin^2\phi\, \gamma[\cos\theta_b + \cos(\theta_t + \phi)]}{R(1 - \cos\phi) + d} \qquad (9.15)$$

This is one part of the pull-off force. The other is composed of the component of the side surface tension on the three-phase lines, whose direction is vertical to the bottom plane. The component of the surface tension along the liquid-bridge axes is $\gamma\sin(\theta_t + \phi)$. The circumference is $2\pi R\sin\phi$. Then, this part of the pull-off force is

$$f_2 = 2\pi R\gamma\sin\phi\sin(\theta_t + \phi) \qquad (9.16)$$

The total pull-off force is

$$f = f_1 + f_2 = \frac{\pi R^2 \sin^2\phi\,\gamma[\cos\theta_b + \cos(\theta_t + \phi)]}{R(1 - \cos\phi) + d}$$
$$+ 2\pi R\gamma\sin\phi\sin(\theta_t + \phi) \qquad (9.17)$$

When the sphere is relatively large (i.e., $R \gg r$ and $\phi \to 0$) and the sphere and the bottom plane are of the same material and are in contact with each other, $d = 0$ and $\theta_t = \theta_b = \theta$, then

$$f = 4\pi R\gamma\sin\theta \qquad (9.18)$$

9.3.3.2 *Capillary condensation resulting in liquid bridge*

1. Capillary condensation of the structure in a system of two parallel plates

Capillary condensation occurs not only in the gap near the spherical-plate contact point but also within other capillary structures. When two parallel plates with hydrophilic surfaces are close to each other in a humid environment, capillary condensation may occur, forming a liquid bridge. The relationship between the curvature radius of the capillary condensation liquid surface and the relative humidity can be described by the Kelvin equation:

$$\left(\frac{1}{r_1} + \frac{1}{r_2}\right)^{-1} = r_k = \frac{\gamma V_{\mathbf{m}}}{\bar{R}T\ln(p_0/p_r)} \tag{9.19}$$

where r_1 and r_2 are the two curvature radiuses of the liquid bridge, $r_{\mathbf{k}}$ is the Kelvin radius, $V_{\mathbf{m}}$ is the mol volume, p_0 is the saturation air pressure of the liquid; for water, at $20°C$, $\gamma V_{\mathbf{m}}/\bar{R}T \approx 0.54\,\text{nm}$. When the liquid bridge is a column, then $r_1 = r$, $r_2 = 0$, and if $p_r/p_0 = 0.9$ (the relative humidity is 90%), then $r_{\mathbf{k}} \approx 5.13\,\text{nm}$. Because the distance between the two parallel plates and the radius of the liquid-bridge surface has the relation $d_0 = 2r_1 \cos\theta \approx 2r_{\mathbf{k}}\cos\theta$, if $\theta = 0$, the distance causing the capillary condensation is $d_0 = 10.23\,\text{nm}$.

2. Capillary condensation of the spherical-plate structure

When the spherical surface contacts the plane, a liquid bridge is naturally formed. If the humidity of the environment is relatively high, capillary condensation may occur, forming a liquid bridge. As the spherical surface and the plane come into contact with each other, the gap between them gradually increases from zero. This gap forms a natural capillary hole of variable scale. Based on the analysis above, when the radius of the spherical surface is very large, the curvature radius of the side surface of the liquid within the capillary, corresponding to different gaps, can be determined as

$$r = \frac{R(1 - \cos\phi)}{2\cos\theta} \tag{9.20}$$

It is also the curvature radius of the liquid surface in the capillary condensation. With ϕ gradually increasing from zero, r gradually increases from zero. Capillary holes are formed with various possible radiuses between the spherical surface and the plane.

According to the Kelvin equation, with a relative humidity of p_r/p_0, the Kelvin radius (average radius) causing the capillary condensation of vapor is

$$r_{\mathbf{k}} = \frac{2\gamma V_{\mathbf{m}}}{\bar{R}T \ln \frac{p_0}{p_r}} \qquad (9.21)$$

Capillary condensation occurs in the zone of $\phi < \phi_r$. Of course, when the actual capillary condensation occurs, ϕ_r may be smaller than the theoretical value, which is mentioned above, due to super-saturation (excessive cold).

9.3.3.3   *Molecular theory model of liquid bridge*

The analysis of liquid cohesive force based on molecular theory is the foundation of the liquid-bridge model within the framework of molecular theory. According to the model of molecular forces, the cohesive work (or interaction potential per unit area) of the liquid can be expressed as

$$W(d) = \frac{A_{11}}{12\pi d^2} \qquad (9.22)$$

where A_{11} is the Hamaker constant and d is the space between the molecules.

The cohesive force (attractive force on unit area) can be written as

$$f(d) = \frac{\partial W(d)}{\partial d} = -\frac{2A_{11}}{12\pi d^3} \qquad (9.23)$$

The surface tension of the liquid can be expressed as

$$\gamma = \frac{W(d)}{2} = \frac{A_{11}}{24\pi d^2} \qquad (9.24)$$

Substituting this surface tension into the equation of cohesive force, it can obtained that

$$f(d) = -\frac{4\gamma}{d} \qquad (9.25)$$

The minus indicates that it is an attractive force.

The weakest position of a liquid bridge is its neck. If it is assumed that the area of its neck, which is vertical to the liquid-bridge axes, is $S_{\mathbf{neck}}$, the stretching resistance of the liquid bridge (pull-off force) can be written as

$$F = S_{\mathbf{neck}} f(d) = -\frac{4\gamma S_{\mathbf{neck}}}{d} \qquad (9.26)$$

9.3.3.4 *Energy theory model of pull-off force*

According to the energy theory of pull-off force, the liquid-bridge system possesses potential energy and adheres to the principle of conservation of work and energy. If there is a very small relative displacement, denoted as δz, between the two solid surfaces, the work done by the pull-off force on this displacement is equal to the variation, denoted as δE, in the potential energy of the liquid-bridge system. The differential expression for this relationship is given by

$$f = \frac{\delta E}{\delta z} \tag{9.27}$$

The key to determining the pull-off force is to determine the total potential energy of the system, especially the potential energy which will change.

9.3.4 *The total free energy of the liquid-bridge*

Surface tension always leads to area contraction. Therefore, as the area increases, the potential energy also increases. Conversely, pressure always leads to volume expansion. Therefore, as the volume decreases, the potential energy increases. The total free energy consists of two parts: the free surface energy and the free bulk energy. The free surface energy arises from the solid–liquid and liquid–vapor interfaces, while the free bulk energy stems from the liquid and vapor volumes under corresponding pressures. In order to calculate the free energy of the liquid-bridge system, one needs to determine both the volume of the liquid and the surface areas of the interfaces. For a liquid bridge, the total free energy can be expressed as

$$E = \sum_{i=1}^{n} \gamma_i A_i - \sum_{j=1}^{m} p_j V_j \tag{9.28}$$

The first sum term on the right-hand side represents the total free surface energy. The second sum term represents the total free bulk energy. Here, A denotes the interfacial area, and γ denotes the interfacial tension. V represents the phase volume, and p represents the phase pressure. The subscripts i and j are used to indicate the interface and the phase, respectively. Naturally, there are usually three phases: the vapor phase, the liquid phase, and the solid

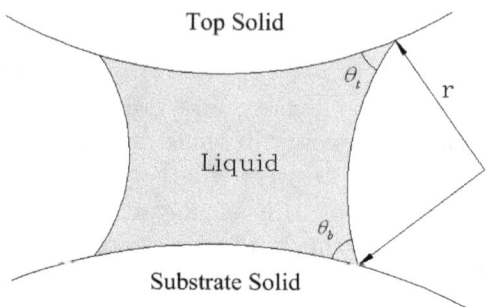

Figure 9.2. A general liquid bridge.

phase. The interface refers to the contact surface between any two of these phases, such as the interface between the vapor phase and the liquid phase, the interface between the vapor phase and the solid phase, and the interface between the liquid phase and the solid phase. A liquid-bridge system is constructed by two solid phases, one liquid phase, and one vapor phase. Therefore, in such a system, there are two vapor–solid interfaces, two liquid–solid interfaces, and one vapor–liquid interface, as shown in Figure 9.2.

If the interface area between the top solid and the liquid is A_{lst} (top wetted area), the interface area between the top solid and the vapor is A_{vst}, the interface area between the substrate solid (bottom solid) and the liquid is A_{lsb} (bottom wetted area), the interface area between the substrate solid and the vapor is A_{vsb}, the interface area between the vapor and the liquid (i.e., the profile of the liquid) is A_{lvp}, the volume of the liquid is V_l, and the volume of the vapor is V_v, the total free energy of the liquid-bridge system, which is described by equation (9.28), can be written as

$$E = \gamma_{\mathsf{lst}} A_{\mathsf{lst}} + \gamma_{\mathsf{vst}} A_{\mathsf{vst}} + \gamma_{\mathsf{lsb}} A_{\mathsf{lsb}} + \gamma_{\mathsf{vsb}} A_{\mathsf{vsb}}$$
$$+ \gamma_{\mathsf{lvp}} A_{\mathsf{lvp}} - p_l V_l - p_{\mathsf{v}} V_{\mathsf{v}} \qquad (9.29)$$

Since the interfacial tension between the solid and the vapor is very small, the effect of the solid–vapor interface in equation (9.29) can be neglected. Then,

$$E = \gamma_{\mathsf{lst}} A_{\mathsf{lst}} + \gamma_{\mathsf{lsb}} A_{\mathsf{lsb}} + \gamma_{\mathsf{lvp}} A_{\mathsf{lvp}} - p_l V_l - p_{\mathsf{v}} V_{\mathsf{v}} \qquad (9.30)$$

When using the energy theory to derive pull-off force, it is important to know the total free energy. At a constant temperature, the differentiation of the total free energy can be derived by

$$\delta E = \gamma_{\text{lst}}\delta A_{\text{lst}} + \gamma_{\text{lsb}}\delta A_{\text{lsb}} + \gamma_{\text{lvp}}\delta A_{\text{lvp}} - p_{\text{l}}\delta \mathbf{V}_{\text{l}} - p_{\text{v}}\delta V_{\text{v}} \quad (9.31)$$

Since the total volume of the liquid and vapor phases is constant, $\delta(V_l + V_v) = 0$, we have

$$\delta V_{\text{v}} = -\delta V_{\text{l}} \quad (9.32)$$

Substituting equation (9.32) into equation (9.31) leads to

$$\delta E = \gamma_{\text{lst}}\delta A_{\text{lst}} + \gamma_{\text{lsb}}\delta A_{\text{lsb}} + \gamma_{\text{lvp}}\delta A_{\text{lvp}} - (p_{\text{l}} - p_{\text{v}})\delta V_{\text{l}} \quad (9.33)$$

For rigid solids, there is no deformation at the interfaces between the solid and the liquid. These interfaces retain the same shapes as the solid surface. However, the interface between the liquid and the vapor will maintain a meniscus shape with corresponding curvatures due to the capillary effect.

9.4 The Solution on the Interface of Liquid and Vapor of the Liquid Bridge

On the meniscus interface of the liquid and the vapor of the liquid bridge, there is no external force, except for the pressure of vapor and the pressure of liquid. According to the energy theory,

$$f_{\text{lvp}} = \frac{\delta E}{\delta r} = \gamma_{\text{lst}}\frac{\delta A_{\text{lst}}}{\delta r} + \gamma_{\text{lsb}}\frac{\delta A_{\text{lsb}}}{\delta r} + \gamma_{\text{lvp}}\frac{\delta A_{\text{lvp}}}{\delta r}$$

$$- (p_{\text{l}} - p_{\text{v}})\frac{\delta V_{\text{l}}}{\delta r} = 0 \quad (9.34)$$

where r is the normal coordinate of the curve meniscus surface. Consider $\delta A_{lst} = \delta A_{lsb} = 0$, then

$$\gamma_{\text{lvp}}\frac{\delta A_{\text{lvp}}}{\delta r} - (p_{\text{l}} - p_{\text{v}})\frac{\delta V_{\text{l}}}{\delta r} = 0 \quad (9.35)$$

If the two principal curvature radiuses of the surface are r_1 and r_2, according to the geometric relationships of the spatial curve surface, we have

$$\delta A_{\mathrm{lvp}} = -\left(\frac{1}{r_1} + \frac{1}{r_2}\right)\delta V_1 \tag{9.36}$$

Substituting equation (9.36) into equation (9.35) leads to

$$\left[\gamma_{\mathrm{lvp}}\left(\frac{1}{r_1} + \frac{1}{r_2}\right) + (p_1 - p_{\mathrm{v}})\right]\frac{\delta V_1}{\delta r} = 0 \tag{9.37}$$

That is,

$$\Delta p = p_{\mathrm{v}} - p_1 = \gamma_{\mathrm{lvp}}\left(\frac{1}{r_1} + \frac{1}{r_2}\right) \tag{9.38}$$

This equation is called the Young–Laplace equation.

9.5   The Pull-Off Force of the Top Solid in the Liquid-Bridge System

9.5.1   *The solution for constant wetted areas*

To obtain the pull-off force of the top solid in the liquid-bridge system, one can use a classical derivative of the energy with respect to the distance between the top solid and the substrate (i.e., the vertical coordinate z). The principle of work–energy conservation can be explained as follows. An incremental displacement δz in the z direction will result in an increase in the total free energy δE of the liquid-bridge system. According to the principle of work–energy conservation, this increase in energy δE should arise from the work done by the pull-off force f_a over the displacement δz.

Then, $\delta E = f_a \cdot \delta z$, and

$$f_a = \frac{\delta E}{\delta z} \tag{9.39}$$

Substituting equation (9.33) into equation (9.39) leads to

$$f_a = \frac{\delta E}{\delta z} = \gamma_{\mathrm{lst}}\frac{\delta A_{\mathrm{lst}}}{\delta z} + \gamma_{\mathrm{lsb}}\frac{\delta A_{\mathrm{lsb}}}{\delta z} + \gamma_{\mathrm{lvp}}\frac{\delta A_{\mathrm{lvp}}}{\delta z}$$
$$- (p_1 - p_{\mathrm{v}})\frac{\delta V_1}{\delta z} \tag{9.40}$$

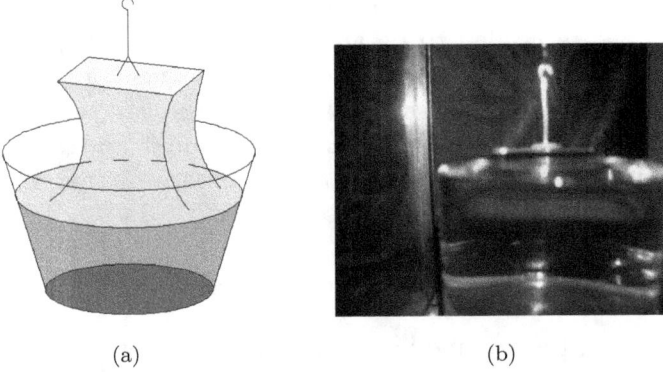

(a) (b)

Figure 9.3. Take liquid with a piece of thin plate from a cup. (a) Scheme and (b) photograph.

For constant interfacial areas, $\delta A_{lst} = \delta A_{lsb} = 0$. Use a piece of thin plate to take the liquid from a cup, as shown in Figure 9.3. The increase of the side area is

$$\delta A_{\mathbf{lvp}} = l_{\mathbf{t}} \cdot \delta z \cdot \sin \theta_{\mathbf{t}} \tag{9.41}$$

where θ_t is the contact angle of the liquid and the top solid, l_t is the perimeter of the interfacial (wetted) area A_{lst}. Whereas $\delta V_l = A_{lst} \cdot \delta z$, and substituting them into equation (9.40) leads to

$$f_{\mathbf{a}} = \frac{\delta E}{\delta z} = \gamma_{\mathbf{lvp}} \cdot l_{\mathbf{t}} \cdot \sin \theta_{\mathbf{t}} - (p_{\mathbf{l}} - p_{\mathbf{v}}) \cdot A_{\mathbf{lst}}$$
$$= \gamma_{\mathbf{lvp}} \cdot l_{\mathbf{t}} \cdot \sin \theta_{\mathbf{t}} + \Delta p \cdot A_{\mathbf{lst}} \tag{9.42}$$

Taking $\gamma_{lvp} = \gamma$ (the surface tension) and $\Delta p = 2\gamma/r_m$, where r_m is the mean radius of the meniscus surface and $2/r_m = 1/r_1 + 1/r_2$, we have

$$f_{\mathbf{a}} = \gamma \cdot l_{\mathbf{t}} \cdot \sin \theta_{\mathbf{t}} + \frac{2\gamma}{r_{\mathbf{m}}} \cdot A_{\mathbf{lst}} \tag{9.43}$$

9.5.2 *An approximate solution for liquid with a constant volume*

In a general liquid-bridge system, the liquid maintains a constant volume. That is, $\delta V_l = 0$. Equation (9.40) can be written as

$$f_{\mathbf{a}} = \frac{\delta E}{\delta z} = \gamma_{\mathbf{lst}} \frac{\delta A_{\mathbf{lst}}}{\delta z} + \gamma_{\mathbf{lsb}} \frac{\delta A_{\mathbf{lsb}}}{\delta z} + \gamma_{\mathbf{lvp}} \frac{\delta A_{\mathbf{lvp}}}{\delta z} \tag{9.44}$$

To obtain the solution of equation (9.44), some geometric relationships are needed to be analyzed and discussed. The depth–width ratio is defined as $\zeta = d/w$. d is the thickness of the liquid film or the minimum width of the gap between the two solid objects. w is the maximum width of area A_{lst}. If the liquid bridge has a small depth–width ratio, $\zeta \ll 1$ and $A_{lst} \approx A_{lsb}$, the volume of the liquid can be approximately expressed as

$$V_l \approx d \cdot A_{lst} \approx d \cdot A_{lsb} \tag{9.45}$$

Considering the volume has a constant value, we get

$$\delta V_l \approx d \cdot \delta A_{lst} + \delta d \cdot A_{lst} = d \cdot \delta A_{lst} + \delta z \cdot A_{lst} = 0 \tag{9.46}$$

Therefore,

$$\frac{\delta A_{lst}}{\delta z} = -\frac{A_{lst}}{d} \tag{9.47}$$

Similarly,

$$\frac{\delta A_{lsb}}{\delta z} = -\frac{A_{lsb}}{d} \tag{9.48}$$

From the geometric relationship for the meniscus surface of the liquid bridge, as shown in Figure 9.4, the relationship between the

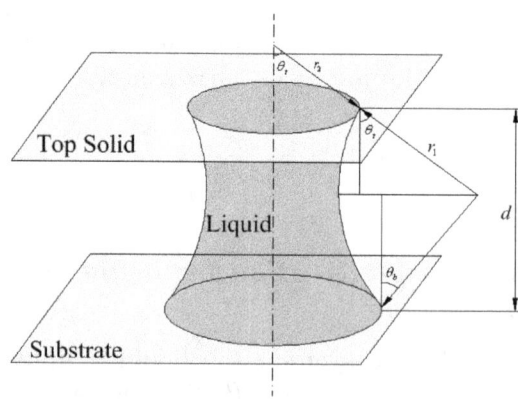

Figure 9.4. A circular liquid bridge.

thickness of the liquid film d (the distance between the two solid objects) and the curvature radius r_1 can be written as

$$d = r_1(\cos\theta_t + \cos\theta_b) \tag{9.49}$$

where θ_b is the contact angle of the liquid and the substrate solid.

Substituting equations (9.41), (9.47), and (9.48) into equation (9.44), since $\gamma_{lst} = -\gamma\cos\theta_t$, $\gamma_{lsb} = -\gamma\cos\theta_b$, and $\gamma_{lvp} = \gamma$, leads to

$$f_a = \frac{\delta E}{\delta z} = \gamma(\cos\theta_t + \cos\theta_b)\frac{A_{lst}}{d} + \gamma\sin\theta_t l_t \tag{9.50}$$

Substituting equation (9.49) into equation (9.50) leads to

$$f_a = \gamma\frac{A_{lst}}{r_1} + \gamma\sin\theta_t l_t \tag{9.51}$$

9.5.3 *An accurate solution for large gap*

If the depth–width ratio is not small, a more general model should be constructed. If the top solid and the substrate solid are both planes, as shown in Figure 9.5, the increment in the vertical distance can be expressed as

$$\delta z = \delta r_1(\cos\theta_t + \cos\theta_b) \tag{9.52}$$

The increment of volume should be written as

$$\delta V_l = -A_{lvp}\delta x + A_{lst}\delta z \tag{9.53}$$

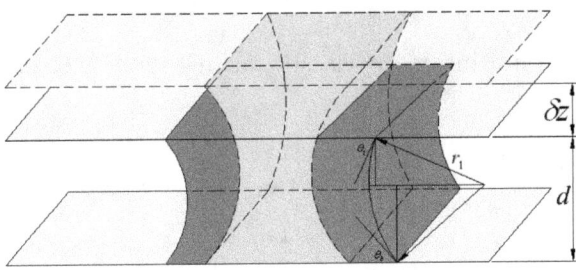

Figure 9.5. A rectangular liquid bridge.

Since the total volume of the liquid maintains a constant value, $\delta V_l = 0$. Therefore,

$$\delta x = \frac{A_{lst}}{A_{lvp}}\delta z \tag{9.54}$$

The increment in the wetted area of the top interface can be written as

$$\delta A_{lst} = -l_t \cdot \delta x = -\frac{l_t A_{lst}}{A_{lvp}}\delta z \tag{9.55}$$

The increment in the wetted area of the substrate interface can be written as

$$\delta A_{lsb} = -l_b \cdot \delta x = -\frac{l_b A_{lsb}}{A_{lvp}}\delta z \tag{9.56}$$

where l_t and l_b are perimeters of the top interface and the substrate interface.

The change in side area can be written as

$$\delta A_{lvp} = l_t \cdot \sin\theta_t \cdot \delta z \tag{9.57}$$

Substituting equations (9.55), (9.56), and (9.57), into equation (9.44), since $\gamma_{lst} = -\gamma\cos\theta_t$, $\gamma_{lsb} = -\gamma\cos\theta_b$, and $\gamma_{lvp} = \gamma$, leads to

$$f_a = \frac{\delta E}{\delta z} = \gamma(l_t\cos\theta_t + l_b\cos\theta_b)\frac{A_{lst}}{A_{lvp}} + \gamma\sin\theta_t l_t \tag{9.58}$$

For a polygonal interface, as shown in Figure 9.6, each side area can be written as

$$A_i = \int_{-(\frac{\pi}{2}-\theta_b)}^{\frac{\pi}{2}-\theta_t} 2\tan\frac{\alpha_i}{2}\left[\frac{l_{ti}}{2}\tan^{-1}\frac{\alpha_i}{2} + r_1\cos\left(\frac{\pi}{2}-\theta\right)_t\right.$$
$$\left. - r_1\cos\beta\right]r_1 d\beta \tag{9.59}$$

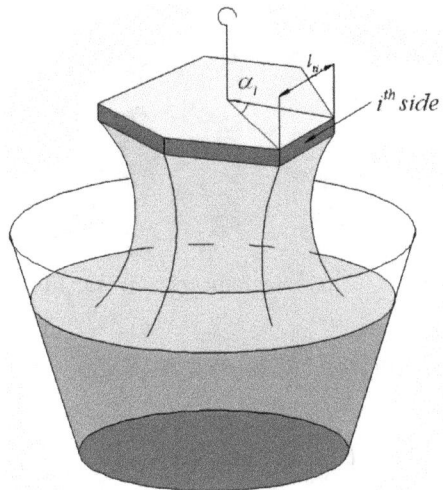

Figure 9.6. A polygonal liquid bridge.

where α_i is the central angle of side i and l_{ti} is the length of side i. Integrating equation (9.32) and summing all the sides leads to

$$A_{\mathbf{lvp}} = \sum_{i=1}^{n} A_i$$

$$= \sum_{i=1}^{n} 2 \tan \frac{\alpha_i}{2} \left[\left(\frac{l_{ti}}{2} \tan^{-1} \frac{\alpha_i}{2} + r_1 \sin \theta_{\mathbf{t}} \right) r_1 (\boldsymbol{\pi} - \theta_{\mathbf{t}} - \theta_{\mathbf{b}}) \right.$$

$$\left. - r_1^2 (\cos \theta_{\mathbf{t}} + \cos \theta_{\mathbf{b}}) \right] \qquad (9.60)$$

For a circular wetted interface, as shown in Figure 9.4, taking the limits $\alpha_i \to 0$ and $n \to \infty$ in (9.60) leads to

$$A_{\mathbf{lvp}} = \lim_{\alpha_i \to 0 (n \to \infty)} \sum_{i=1}^{n} A_i$$

$$= \int_0^{2\boldsymbol{\pi}} [(r_2 \sin \theta_{\mathbf{t}} + r_1 \sin \theta_{\mathbf{t}}) r_1 (\boldsymbol{\pi} - \theta_{\mathbf{t}} - \theta_{\mathbf{b}})$$

$$- r_1^2 (\cos \theta_{\mathbf{t}} + \cos \theta_{\mathbf{b}})] d\alpha$$

$$= 2\boldsymbol{\pi} r_1 [(r_2 + r_1) \sin \theta_{\mathbf{t}} (\boldsymbol{\pi} - \theta_{\mathbf{t}} - \theta_{\mathbf{b}}) - r_1 (\cos \theta_{\mathbf{t}} + \cos \theta_{\mathbf{b}})]$$

$$(9.61)$$

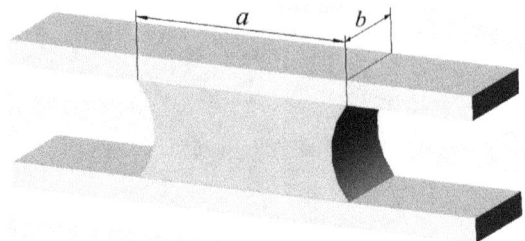

Figure 9.7.   A narrow, long rectangular liquid bridge.

For a rectangular interface, as shown in Figure 9.7, the side area is

$$
A_{\mathbf{lvp}} = \sum_{i=1}^{n} A_i = 4\left\{ \frac{b}{a} \left[\left(\frac{a}{2} + r_1 \sin \theta_{\mathbf{t}} \right) r_1 (\pi - \theta_{\mathbf{t}} - \theta_{\mathbf{b}}) \right.\right.
$$

$$
\left. - r_1^2 (\cos \theta_{\mathbf{t}} + \cos \theta_{\mathbf{b}}) \right] + \frac{a}{b} \left[\left(\frac{b}{2} + r_1 \sin \theta_{\mathbf{t}} \right) r_1 (\pi - \theta_{\mathbf{t}} - \theta_{\mathbf{b}}) \right.
$$

$$
\left.\left. - r_1^2 (\cos \theta_{\mathbf{t}} + \cos \theta_{\mathbf{b}}) \right] \right\} \tag{9.62}
$$

where a and b are the length and width of the rectangle, respectively. For a square interface,

$$
A_{\mathbf{lvp}} = \sum_{i=1}^{n} A_i
$$

$$
= \left[(l_{\mathbf{t}} + 8r_1 \sin \theta_{\mathbf{t}}) r_1 (\pi - \theta_{\mathbf{t}} - \theta_{\mathbf{b}}) - 8r_1^2 (\cos \theta_{\mathbf{t}} + \cos \theta_{\mathbf{b}}) \right] \tag{9.63}
$$

Here, for a narrow, long rectangular solid, as shown in Figure 9.7, the interface area will contract from the long side to form a square shape and then contract from the short side to form a circular shape. Substituting equation (9.61) into equation (9.58), leads to

$$
f_{\mathbf{a}} = \gamma \frac{(l_{\mathbf{t}} \cos \theta_{\mathbf{t}} + l_{\mathbf{b}} \cos \theta_{\mathbf{b}}) A_{\mathbf{lst}}}{r_1 (l_{\mathbf{t}} + 2\pi r_1 \sin \theta_{\mathbf{t}})(\pi - \theta_{\mathbf{t}} - \theta_{\mathbf{b}}) - 2\pi r_1 (\cos \theta_{\mathbf{t}} + \cos \theta_{\mathbf{b}})}
$$

$$
+ \gamma \sin \theta_{\mathbf{t}} l_{\mathbf{t}} \tag{9.64}
$$

If the material of the top solid is the same as that of the substrate solid, $\theta_t = \theta_b$ and $l_t = l_b$,

$$f_{\mathbf{a}} = \gamma \frac{2l_{\mathbf{t}} \cos \theta_{\mathbf{t}} A_{\mathrm{lst}}}{r_1(l_{\mathbf{t}} + 2\pi r_1 \sin \theta_{\mathbf{t}})(\pi - 2\theta_{\mathbf{t}}) - 4\pi r_1 \cos \theta_{\mathbf{t}}} + \gamma \sin \theta_{\mathbf{t}} l_{\mathbf{t}}$$

$$(9.65)$$

For a liquid bridge with a small depth–width ratio, $l_t. \gg 2r_1 \cos \theta_t$. Equation (9.65) can be written as

$$f_{\mathbf{a}} = \gamma \frac{2 \cos \theta_t A_{\mathrm{lst}}}{r_1(\pi - 2\theta_t)} + \gamma \sin \theta_t l_{\mathbf{t}} \qquad (9.66)$$

When comparing equation (9.51) with equation (9.66), it can be observed that these two equations are not identical. Equation (9.66) is a more precise equation, as it takes into account the curved surface of the sides during the calculation of the liquid volume. Using the straight line distance $2r_1 \cos \theta_t$ instead of the curve arc distance $r_1(\pi - 2\theta_t)$, equation (9.66) will change to equation (9.51). For a square interface, substituting equation (9.63) into equation (9.58) and considering $l_t. \gg 2r_1 \cos \theta_t$, the same equation as (9.66) can be obtained.

It must be pointed out that, as the solid–liquid interface contracts, with its border moving away from every boundary, the shape of the interface tends to become a circle due to surface tension.

9.5.4 An accurate solution for the sphere-plane liquid-bridge system with a large gap

When the top solid has a spherical shape and the substrate solid is planar, as shown in Figure 9.8(a), the situation becomes relatively complicated. The conservation condition of volume can also be written as

$$\delta V_{\mathrm{l}} = -A_{\mathrm{lvp}}\delta x + A_{\mathrm{lst}}\delta z = 0 \qquad (9.67)$$

Then,

$$\delta x = \frac{A_{\mathrm{lst}}}{A_{\mathrm{lvp}}}\delta z \qquad (9.68)$$

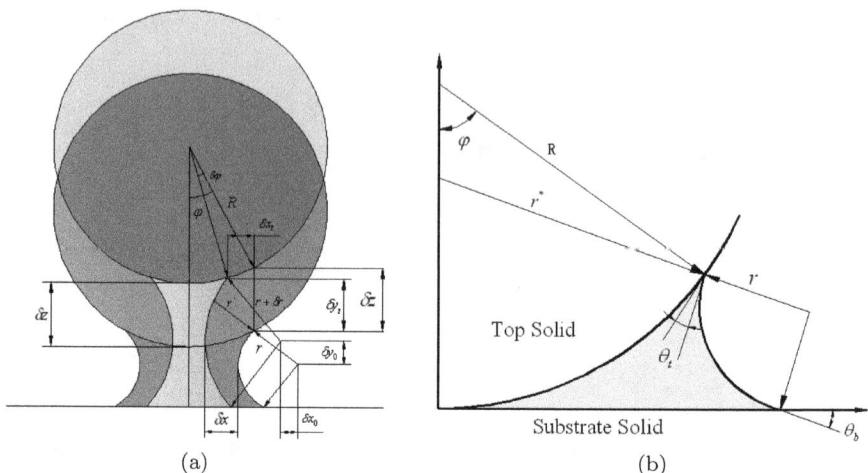

(a)   (b)

Figure 9.8. A sphere–plane liquid bridge system with a large gap. (a) Different gaps of liquid bridge and (b) geometric relationship between sphere and plane.

When the sphere moves upward with a slight displacement δz, the meniscus interface between the liquid and the vapor, the curvature center, and the curvature radius will also undergo changes. However, the contact angles should remain constant. Assuming that the curvature center experiences a slight displacement to the left by δx_0 and a slight upward displacement by δy_0 and that the curvature radius increases by δr, the wetted boundary on the spherical interface will contract with a displacement to the left by δx_t, which can be expressed as

$$\delta x_{\mathbf{t}} = \delta x_0 + \delta r \sin(\theta_{\mathbf{t}} + \phi) - r \cos(\theta_{\mathbf{t}} + \phi)\delta\phi \qquad (9.69)$$

where ϕ is called the filling angle and $\delta\phi$ is defined as the positive in clockwise direction.

The vertical upward displacement can be written as

$$\delta y_{\mathbf{t}} = \delta y_0 + \delta r \cos(\theta_{\mathbf{t}} + \phi) + r \sin(\theta_{\mathbf{t}} + \phi)\delta\phi \qquad (9.70)$$

For the geometric relationships, as shown in Figure 9.8, the displacement of the meniscus curve surface (or the displacement in the left-hand direction of the horizontal apex of the profile) can be written as

$$\delta x = \delta x_0 + \delta r \qquad (9.71)$$

The wetted area of the substrate plane will contract with a following decrement:

$$\delta\rho = \delta x - (1 - \sin\theta_b)\delta r \qquad (9.72)$$

Since $\delta x_t = R\cos\phi\,\delta\phi$ and $\delta y_0 = \delta r\cos\theta_b$, we can obtain the following relationships:

$$\delta z = \delta y_t + R\sin\phi\,\delta\phi \qquad (9.73)$$

and

$$\delta y_t = [\cos\theta_b + \cos(\theta_t + \phi)]\delta r + r\sin(\theta_t + \phi)\delta\phi \qquad (9.74)$$

Substituting equation (9.74) into equation (9.73) leads to

$$\delta z = [\cos\theta_b + \cos(\theta_t + \phi)]\delta r + [R\sin\phi + r\sin(\theta_t + \phi)]\delta\phi \qquad (9.75)$$

Substituting equation (9.69) into equation (9.71) leads to

$$\delta x = [1 - \sin(\theta_t + \phi)]\delta r + [R\cos\phi + r\cos(\theta_t + \phi)]\delta\phi \qquad (9.76)$$

From equations (9.68), (9.72), (9.74), and (9.75), the following solutions can be obtained:

$$\frac{\delta r}{\delta z} = \frac{C_1 - C_3\xi}{C_1 C_2 - C_3 C_4} \qquad (9.77)$$

$$\frac{\delta\phi}{\delta z} = \frac{C_2\xi - C_4}{C_1 C_2 - C_3 C_4} \qquad (9.78)$$

$$\frac{\delta\rho}{\delta z} = \xi - \frac{C_5(C_1 - C_3\xi)}{C_1 C_2 - C_3 C_4} \qquad (9.79)$$

where $C_1 = R\cos\phi + r\cos(\theta_t + \phi)$, $C_2 = \cos\theta_b + \cos(\theta_t + \phi)$, $C_3 = R\sin\phi + r\sin(\theta_t + \phi)$, $C_4 = 1 - \sin(\theta_t + \phi)$, $C_5 = 1 - \sin\theta_b$, and $\xi = \frac{A_{lst}}{A_{lvp}}$.

The wetted area on the sphere can be written as

$$A_{lst} = 2\pi R^2(1 - \cos\phi) \qquad (9.80)$$

The wetted area on the substrate plane can be written as

$$A_{\mathsf{lsb}} = \pi \rho^2 \tag{9.81}$$

The increment of meniscus area can be described as

$$\delta A_{\mathsf{lvp}} = 2\pi R \sin \phi \delta s \tag{9.82}$$

where δs is the increment of the meridian of the profile.

$$\delta s = r\delta\phi + \delta r(\pi - \theta_{\mathsf{t}} - \theta_{\mathsf{b}} - \phi) \tag{9.83}$$

With the relationships given in equations (9.77–9.79), deriving equations (9.80–9.82) with respect to the vertical coordinate z leads to

$$\frac{\delta A_{\mathsf{lst}}}{\delta z} = -\frac{2\pi R^2 \sin \phi (C_2 \xi - C_4)}{C_1 C_2 - C_3 C_4} \tag{9.84}$$

$$\frac{\delta A_{\mathsf{lsb}}}{\delta z} = -2\pi \rho \left[\xi - \frac{C_5 (C_1 - C_3 \xi)}{C_1 C_2 - C_3 C_4} \right] \tag{9.85}$$

$$\frac{\delta A_{\mathsf{lvp}}}{\delta z} = 2\pi R \sin \phi \frac{r(C_2 \xi - C_4) + (C_1 - C_3 \xi)(\pi - \theta_{\mathsf{t}} - \theta_{\mathsf{b}} - \phi)}{C_1 C_2 - C_3 C_4} \tag{9.86}$$

Substituting equations (9.84–9.86) into equation (9.44), since $\gamma_{\mathsf{lst}} = -\gamma \cos \theta_{\mathsf{t}}$, $\gamma_{\mathsf{lsb}} = -\gamma \cos \theta_{\mathsf{b}}$, and $\gamma_{\mathsf{lvp}} = \gamma$, leads to

$$\begin{aligned} f_{\mathsf{a}} = {} & \frac{2\pi R^2 \sin \phi (C_2 \xi - C_4)}{C_1 C_2 - C_3 C_4} \gamma \cos \theta_{\mathsf{t}} \\ & + 2\pi \rho \left[\xi - \frac{C_5 (C_1 - C_3 \xi)}{C_1 C_2 - C_3 C_4} \right] \gamma \cos \theta_{\mathsf{b}} \\ & + 2\pi R \sin \phi \frac{r(C_2 \xi - C_4) + (C_1 - C_3 \xi)(\pi - \theta_{\mathsf{t}} - \theta_{\mathsf{b}} - \phi)}{C_1 C_2 - C_3 C_4} \gamma \end{aligned} \tag{9.87}$$

When ϕ tends to zero (i.e., $\phi \to 0$), the liquid-bridge force f_a also approaches zero and does not maintain a constant value. This is consistent with practical physical phenomena. However, if an approximate cylindrical side surface is used instead of an accurate meniscus profile to calculate the side area or the volume of liquid, a conflict will arise. This issue can be addressed as follows.

When the thickness of the liquid film is very small, the curvature radius will be much smaller than the radius of the sphere ($r \ll R$). In this scenario, instead of using the actual meniscus surface, an approximate cylindrical surface can be employed. The side area described by equation (9.61) (where $r_1 \ll r_2$) can be approximated as

$$A_{\mathbf{lvp}} = 2\pi R \sin \phi (D + d) \tag{9.88}$$

where D is the distance between the sphere and the plane and d is the height of the spherical cap.

Neglecting the profile surface tension (the third term in equation (9.87)), considering $\theta_b = \theta_t = \theta$ and $\rho = R \sin \phi$, and taking $D = 0$ and the limit of $\phi \to 0$, then substituting equations (9.88) and (9.80) into equation (9.87) leads to

$$f_{\mathbf{a}} = 4\pi R \gamma \cos \theta \tag{9.89}$$

This is the conventional equation. Since equation (9.89) is independent of the filling angle ϕ, this expression shows that the liquid-bridge force approaches a constant value when ϕ tends to zero. Of course, this is in conflict with the fact that there is no pull-off force between two completely dry surfaces.

The wetted area on the plane can be approximately written as

$$A_{\mathbf{lsb}} = \pi R^2 \sin^2 \phi \tag{9.90}$$

Equation (9.77) can be approximately written as

$$\delta x = R \cos \phi \delta \phi \tag{9.91}$$

The conservation condition for the volume approximately results in

$$\delta x = \frac{A_{\mathbf{lsb}}}{A_{\mathbf{lvp}}} \delta z \tag{9.92}$$

Substituting equations (9.88), (9.90), and (9.91) into equation (9.92) leads to

$$\frac{\delta \phi}{\delta z} = \frac{\tan \phi}{2(D + d)} = \frac{\tan \phi}{2R(1 - \cos \phi)(1 + D/d)} \tag{9.93}$$

Taking the derivatives of equations (9.80), (9.82), and (9.90) with respect to the vertical coordinate z and substituting equation (9.93) into them leads to

$$\frac{\delta A_{\text{lst}}}{\delta z} = -\frac{2\pi R^2 \sin\phi \tan\phi}{2R(1 - \cos\phi)(1 + D/d)} \tag{9.94}$$

$$\frac{\delta A_{\text{lsb}}}{\delta z} = -\pi R^2 2 \sin\phi \cos\phi \frac{\tan\phi}{2R(1 - \cos\phi)(1 + D/d)} \tag{9.95}$$

$$\frac{\delta A_{\text{lvp}}}{\delta z} = 2\pi R \sin\phi \tag{9.96}$$

Substituting equations (9.94–9.96) into equation (9.44) and using $\gamma_{\text{lst}} = -\gamma \cos\theta$, $\gamma_{\text{lsb}} = -\gamma \cos\theta$, and $\gamma_{\text{lvp}} = \gamma$ leads to

$$f_{\mathbf{a}} = \pi R \gamma \cos\theta \frac{(1 + \cos\phi)^2}{\cos\phi(1 + D/d)} + 2\pi R \gamma \sin\phi \tag{9.97}$$

If the surface tension is negligible, equation (9.97) can be written as

$$f_{\mathbf{a}} = \pi R \gamma \cos\theta \frac{(1 + \cos\phi)^2}{\cos\phi(1 + D/d)} \tag{9.98}$$

Although equation (9.98) is not independent of the filling angle ϕ, the same form as equation (9.89) can be obtained. This is because an approximate cylindrical surface is also used for the calculation of the liquid volume. As $D = 0$ (Figure 9.8(b)) and $\phi \to 0$, equation (9.98) can be changed into

$$f_{\mathbf{a}} = 4\pi R \gamma \cos\theta \tag{9.99}$$

This is the same as the conventional equation (9.89).

It should be pointed out that equation (9.99) is only applicable for large spheres and thin liquid films. Although equation (9.98) partially modifies equation (9.99), it is only applicable for thin liquid films with a small depth–width ratio ($\zeta \ll 1$).

The model built in this section is applicable not only to large spheres and thin liquid films but also to small spheres and large separating distances.

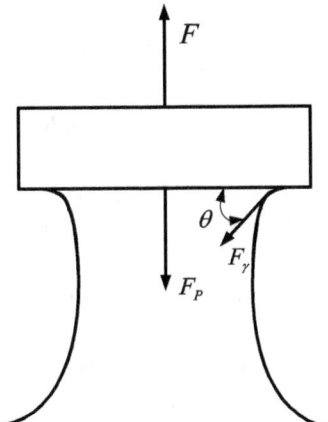

Figure 9.9. Diagram of force before wetting the edge.

9.5.5 *The edge-effect of liquid-bridge*

In general, a liquid-bridge system consists of a top solid, a bottom solid, and the liquid situated between them. The liquid not only wets the major surface of the top solid but also the profile in the vicinity of the solid's edge, as illustrated in Figure 9.9. In this case, the pull-off force will be caused not only by the negative pressure F_P but also by the profile surface tension of liquid, F_γ:

$$F = F_P + F_\gamma \sin\theta \tag{9.100}$$

where $F_P = \Delta p A$ is the force caused by the pressure Δp of the meniscus surface of the liquid. A represents the wetted area of the top solid. Δp is determined using the Laplace equation. γ is the surface tension of the liquid. r_1 and r_2 are the principal curvature radiuses of the liquid side surface on the plane, which are vertical and parallel to the liquid bridge. If r is the average curvature radius of the meniscus surface, then $2/r = 1/r_1 + 1/r_2$. Substituting it into the Laplace equation leads to

$$\Delta p = \frac{2\gamma}{r} \tag{9.101}$$

Substituting (9.101) into $F_P = \Delta p A$ leads to

$$F_P = \frac{2\gamma}{r} A \tag{9.102}$$

The force caused by surface tension of the liquid can be written as $F_\gamma = l\gamma$. Then, expression (9.100) can be written as

$$F = \frac{2\gamma}{r}A + l\gamma \sin\theta \qquad (9.103)$$

where θ is the contact angle of the liquid with respect to the top solid and l is the perimeter of the wetted area.

Actually, in most liquid-bridge systems, the liquid not only wets the lower surface of the top solid but also its edge. Since the top solid in a liquid bridge has some thickness, both the bottom surface and the side surfaces of the top solid are wetted, as shown in Figure 9.10. This will cause a force on the top solid within the liquid bridge. Therefore, the edge effect of the liquid bridge needs to be considered when researching the pull-off force of the top solid.

For analytical convenience, suppose the top solid has a certain thickness (such that its profile cannot be ignored), which shares the same characteristics as the lower surface of the top solid, including the same material, smoothness, contact angle, and surface tension. If only the profile of the top solid is wetted and the other surfaces do not come into contact with the liquid (as illustrated in Figure 9.11), then the component of the surface tension in the direction parallel

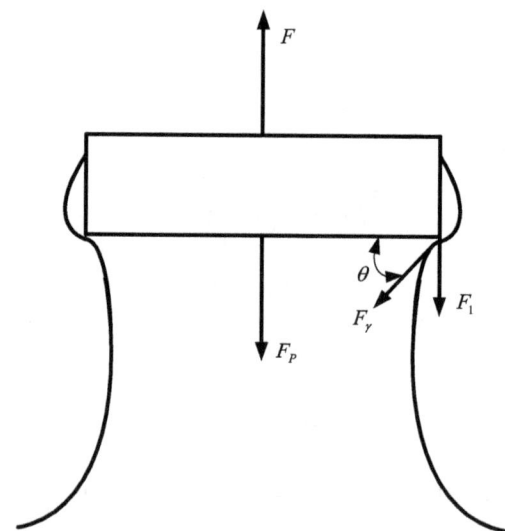

Figure 9.10. Actual wetting of liquid bridge.

to the profile of the top solid is

$$F_1 = F_A \cos\theta = l\gamma\cos\theta \qquad (9.104)$$

In an actual liquid-bridge system, if the liquid wets the top solid up to its edge, it will wet both the lower surface and the profile of the top solid. Therefore, the actual situation is not as depicted in Figure 9.9 but as in Figure 9.10. In this case, the pull-off force acting on the top solid in the liquid bridge includes two components: one is caused by the wetting of the lower surface of the top solid (as shown in Figure 9.9); the other is caused by the wetting of the profile of the top solid (as shown in Figure 9.11). Therefore, the total pull-off force acting on the solid is the sum of the forces depicted in Figures 9.9 and 9.10, as illustrated in Figure 9.12:

$$F = F_P + F_\gamma \sin\theta + F_1 \qquad (9.105)$$

or

$$F = \frac{2\gamma}{r}A + l\gamma\sin\theta + l\gamma\cos\theta \qquad (9.106)$$

Expression (9.106) is the formula used for calculating the pull-off force when considering the edge effect. Some experiments have been

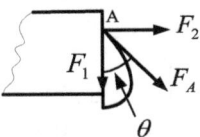

Figure 9.11. Force diagram of wetting the profile of top solid.

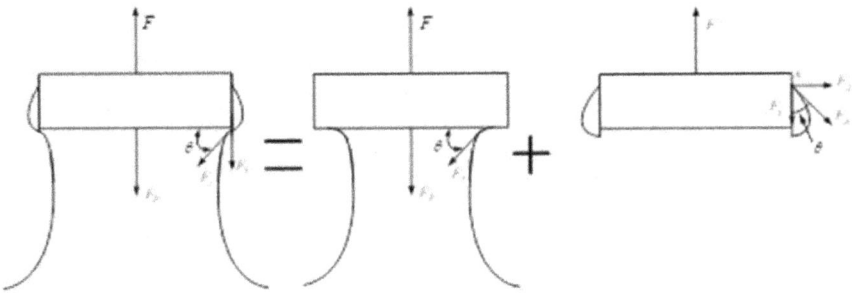

Figure 9.12. Diagram for the calculating model of the pull-off force considering the edge effect.

Table 9.1. Several results of calculated pull-off force and the measured pull-off force (unit of force: mN; unit of scale of silicon film: mm).

No.	Scale of silicon film	F_P	$F_\gamma \sin \theta$	F_2	F_1	F	Measurement
1	15.71 × 15.71	5.044	4.218	9.262	1.755	11.017	11.84
2	30 × 10	6.624	5.373	11.997	2.235	14.232	14.67
3	20 × 20	7.805	5.373	13.178	2.235	15.413	16.45
4	10 × 20	4.195	4.03	8.225	1.677	9.902	9.13
5	19.64 × 19.64	7.894	5.277	13.171	2.195	15.366	17.62
6	11.4 × 20	4.666	4.218	8.884	1.755	10.639	10.16
7	31.2 × 10	6.545	5.534	12.079	2.302	14.381	14.4
8	17.73 × 17.73	6.443	4.763	11.196	1.982	13.178	13.86
9	15.71 × 20	6.591	4.797	11.388	1.996	13.384	13.76

Remark: $F_2 = F_P + F_\gamma \sin \theta$ is the pull-off force without considering the edge effect.

$F = F_P + F_\gamma \sin \theta + F_1$ is the pull-off force considering the edge effect.

$F_1 = F_A \cos \theta = l\gamma \cos \theta$ the force caused by the edge effect.

done to measure the pull-off force and prove the validity of expression (9.106).

In the measurement, the contact angle of silicon and water is 67.412° at 18°C, and the surface tension of water is 72.746 mN/m at 17°C. A series of data points was obtained by measuring water wetting silicon films of various scales. The calculated data were compared with the measured data in Table 9.1. As shown in Table 9.1, the pull-off force calculated using expression (9.106) is closer to the measured value. Therefore, the model that considers the edge effect is valid and reasonable.

Chapter 10

MEMS and Capillary Action

Science and technology are continually advancing toward miniaturization, with scales reaching the micrometer and nanometer levels. In terms of machining accuracy and structural dimensions, micrometer-level technology has become increasingly sophisticated. Simultaneously, nanoscale processing technology and functional devices are evolving, enabling researchers to explore atomic and molecular scales within the realm of micro-mechanical processing technology. Capillarity is observed at both macroscopic and microscopic scales. However, capillary action is particularly pronounced in Micro-Electro-Mechanical Systems (MEMS) due to the relatively small volume of the devices, their relatively large surface area, and the minimal distance between adjacent surfaces. With the rapid development of MEMS technology, micro-electro-mechanical devices have gained widespread application, leading to increased attention on capillarity within MEMS.

10.1 Capillary Action in the Processing of MEMS

The processes of producing MEMS are diverse; there are more than 110 different kinds of processes. However, only seven kinds of processes are specially applied for some products or in some companies. All of the other kinds can be easily divided into standardized processes. Releasing the suspended structure is a common process of producing a MEMS device. In order to make a suspended structure, some appropriate materials, a structural layer, and an etching process of removing the materials around the structure are needed. If the removed materials come from the substrate, it is called Bulk MicroMachining (BMM). Otherwise, when the removed materials are deposited, it is called Surface MicroMachining (SMM). For example, the pressure sensors are mainly fabricated using BMM, the products of RF MEMS are fabricated through the SMM of metals, and microphones utilize the SMM technology based on dielectric materials. If you want to use MEMS technology to process a micropump, the multi-wafer stacked structure processed by BMM is the first choice. The products machined by BMM have some advantages, such as high accuracy and robust structure. However, the cost is relatively high, and it is difficult to integrate them with CMOS. The development trend of the process of producing MEMS is to integrate with CMOS and develop from BMM to SMM.

During the process of producing and packaging MEMS, capillarity will have a great influence on the properties and reliability of micromachines. The capillary action is important for micro-devices and MEMS structures. In recent years, there have been more and more studies on capillary mechanics in microscale.

10.1.1   *Capillary adhesion in MEMS*

Many factors influence the phenomenon of micro-machine adhesion, mainly due to the role of surface tension, such as capillary gravitation, van der Waals force, hydrogen bonds, and static electricity. The strengths of these kinds of surface tension effects are influenced by many factors, including the material of the structure, the surface complexion of the material, relative movement, humidity, and temperature. This section only discusses the influence of capillary adhesion.

Etching is used in MEMS to chemically remove the layers, which are called "sacrificial layers", from the surface of a wafer during manufacturing. Etching is a very important process module. Every wafer undergoes many etching steps before it is completed. The etching technique can be divided into wet etching and dry etching. The process of wet etching uses liquid-phase ("wet") etchants. Dry etching is a plasma etching method, where the process involves a physical action using plasma to attack the wafer surface. Once the sacrificial

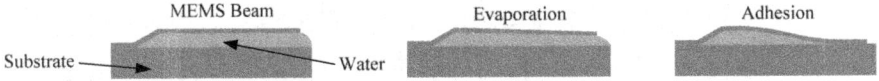

Figure 10.1. Adhesion caused by water cleaning.

layer is removed, the activity structure of micro-devices can move freely as designed. Upon release, the device is easy to attack and influence. It will be destroyed by mechanical vibration or be stained by pollution.

For most MEMS devices, etching is the most important procedure because the sacrificial layer must be removed under accurate control. After release, the structure must be entirely cleaned to remove etchants, dissolvable materials, solid remains, and other by-products. Adhesion between the movement parts of micro-structures is a serious problem. When the mechanical restitution force of micro-structures is smaller than the surface adhesion force, adhesion occurs. Water-based cleaners can promote adhesion. The liquid will form a liquid meniscus on the hydrophilic surface when the device is taken away from the water solution after wet etching. After evaporation and contraction, the micro-structure will lean onto the substrate. Finally, the adhesion forms, as shown in Figure 10.1.

Capillary adhesion exists in traditional bulk micromachining, as shown in Figure 10.2. The process includes deposition, dry etching, anodic bonding, release, etc. Compared to the SMM process, this technique can produce devices with a high depth–width ratio, which is highly attractive for devices based on comb drivers and comb capacitance sensors. In addition, the silicon deposited with dense boron can be directly used as an electrode, which makes the process easy. Firstly, the silicon film is deposited with dense boron (Figure 10.2(a)). Secondly, photolithography and dry etching are used to create the bonding table (Figure 10.2(b)). Then, the whole structure is made using the second photolithography (see Figure 10.2(c)).

The electrode of titanium, platinum, and gold is made on Pyrex 7740 glass through the usual process. Grooves must be etched into the glass before sputtering so that it ensures that the electrode surface lies 50 nm higher than the glass surface (Figure 10.2(d–f)). The silicon film and the glass film are cleaned by ultrasonic waves to reduce the pollution by solid particles and immersed in ammonia to increase their hydrophilicity. Then, the two films are aligned and

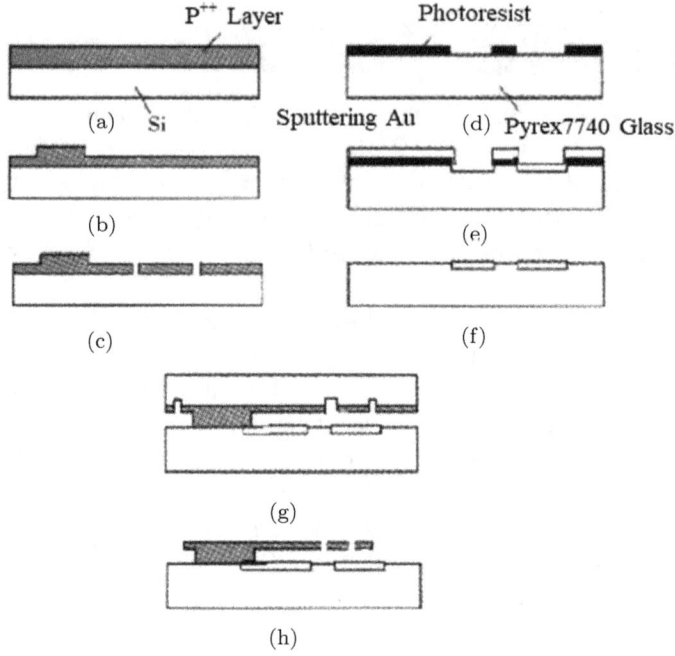

Figure 10.2. Process diagram of traditional bulk micromachining.

pasted. After higher-temperature and electrostatic processing, the silicon and the glass are bonded together. Then, they are placed in an Ethylenediamine, o-Phthalic acid, and Water (EPW) or potassium hydroxide solution to erode the low-density deposits of silicon. Finally, after soaking in methanol, they are directly placed in an oven to release the active structure (Figure 10.2(h)). The capillary force caused by the EPW or potassium hydroxide solution is large enough to pull the movable plate toward the substrate.

Besides the traditional bulk silicon-dissolved wafer process, SMM is the other important technique in MEMS processes. This process is considered one of the key techniques in the production of MEMS devices. Firstly, a thin film is deposited onto the substrate. Then, the film is selectively etched to form the designed structure. At present, some products manufactured by this MEMS technique have been widely used, such as pressure sensors, acceleration sensors, optical switches, and Digital Mirror Displays (DMDs). The main advantages of SMM are as follows: it is compatible with traditional and mature

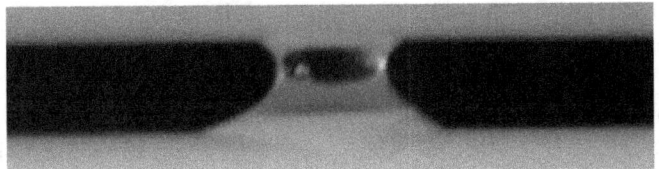

Figure 10.3.   Liquid bridge between micro parallel plates.

Integrated Circuit (IC) fabrication techniques, and the devices manufactured by surface processing are easily integrable with control circuits.

For SMM, after the sacrificial layer is etched, the device needs to be cleaned with deionized water to remove the etchant and any etched materials. When the device is taken out of the deionized water, a liquid bridge is formed between the two parallel plates, as shown in Figure 10.3. The capillary force caused by the liquid bridge always appears during the drying process after wet etching.

When the liquid come into contact with the adjacent surface and slowly evaporates, the capillary force caused by the liquid cleanser can pull the parts together. When the capillary force is large enough to cause the adhesion, the liquid evaporation will cause a more serious problem. The volume of the liquid is greatly reduced during evaporation, which produces a large enough force to cause the collapse of fragile suspension structures. The shrinking droplet pulls the MEMS surface closer and closer until they touch each other and stick.

At the last step of micromachining, the micro-structure must be rinsed and dehydrated. At the end of dehydration, the liquid disappears from the chip surface. However, liquid still exists in the gap of the micro-structure. Liquid bridges of different curvatures are formed on the surface of the residual liquid. With the effect of surface tension, the liquid bridge force pulls the microstructure to the substrate, which will destroy the structure or cause the moveable structure to stick on the substrate forever.

10.1.2   Control of capillary adhesion

Capillary adhesion includes several important factors which influence the performance of MEMS devices, such as yield rate, service life, and reliability. These increase the scrap rate of MEMS devices and lead to unnecessary losses. Therefore, it is important to effectively control capillary adhesion in MEMS devices during their processing and manufacturing.

Methods of dry etching, using non-water cleaning or drying with supercritical carbon dioxide, can reduce or even eliminate the adhesion during the releasing process. Supercritical drying is a common method to reduce capillary force. MEMS devices are dried in liquid dioxide carbon, which makes them reach the supercritical point by adjusting the pressure and temperature. During the process of drying, the meniscus cannot be formed and the surfaces cannot be pulled together. Of course, the various parts can also be chemically treated to ensure that the surfaces of the devices do not stick even when they come into contact with each other. The extraction by supercritical carbon dioxide can remove the solvent and release the complex and fragile MEMS structure on the silicon. The structures which are released and cleaned include micro-engines with a single gear, bridges and cantilevers, pressure sensors, and comb drivers. The surface tension of the supercritical liquid can be ignored. The liquid can also remove the solvent in the capillary gap of only a few nanometers. In addition, the device can be dried by supercritical carbon dioxide repeatedly without damaging its structure.

The traditional bulk silicon-dissolved wafer process can also be improved to better control capillary adhesion. The new bulk silicon-dissolved wafer process is shown in Figure 10.4. Firstly, the silicon film is deposited with P^{++} and a 2 μm bonding table is created using photolithography and dry etching. At the same time, the bonding frame is manufactured around each chip to surround it. The electrode is formed on Pyrex 7400 glass through lithography and sputtering. Then, the silicon and glass are aligned and electrostatically bonded together. After bonding, the film is put into an EPW solution to erode the low-density silicon. The liquid could not flow into the single chip due to the sealed frames. Then, the final activating structure is formed by double-side lithography and dry etching.

It is obvious that this kind of process can avoid residual liquid formed in the traditional process so that it can reduce adhesion. In addition, in this new process, the liquid bridge force does not occur because the activating structure does not come into contact with the liquid. Therefore, this technology increases the rate of qualified products produced by the devices.

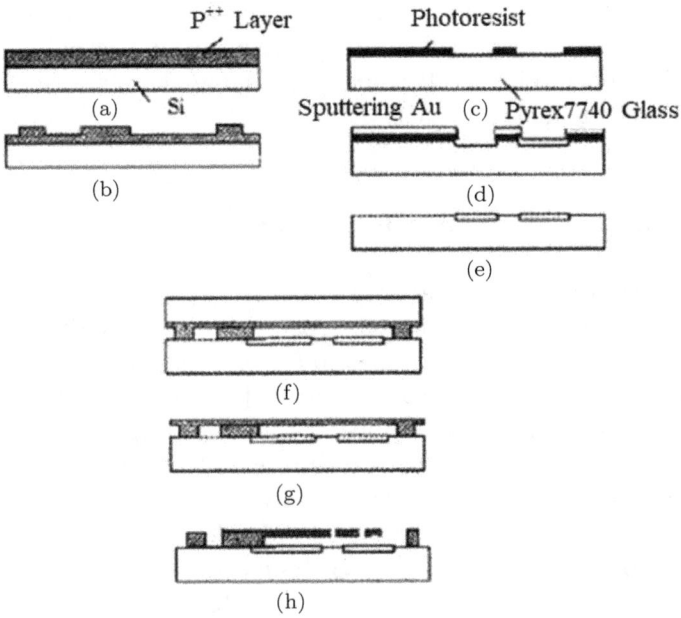

Figure 10.4. New BMM process.

The performance of MEMS devices is affected by their internal environment. Additives can effectively improve the internal environment of MEMS devices. An additive is necessary for some MEMS devices that are very sensitive to humidity and small pollution particles. Even extremely small amounts of liquid or nanoparticles can cause the failure of the device. Adding an additive during or before sealing MEMS devices can effectively control the adhesion. Adsorbents are the most important additives, which can optionally clear small amounts of liquid and nanoparticles produced during the manufacture of MEMS devices. Some adsorbents can also absorb some other kinds of harmful materials in the package. Adsorbents can be divided into three types: gas, liquid, and solid. Hydrogen and oxygen molecules have been found in airtight packages of MEMS devices, which is known to have potential hazards. The gas adsorbent can effectively eliminate the hazard. Water is the main substance to be absorbed by a liquid adsorbent, which is generally called a wet adsorbent. Micro- or nano-sized solid adsorbents are more generally used, which means that they can capture any small particles of solid components. Adsorbents were first used in vacuum tubes to absorb oxygen molecules which may lead to performance degradation of glow wires. At present, adsorbents are mostly used for packaging satellite communication modules. Of course, the same type of adsorbent can also be used for packaging MEMS.

10.2   Capillary-Driven MEMS Fluid Self-Assembly

The self-assembly technique is a process using microscopic forces, such as surface tension and electrostatic force, as driving forces to make a large number of micro-devices automatically arrange (location and orientation) from disorder to order according to some setting rules and then couple and fix them. During this process, the capillary-driven self-assembly is an important method. This concept originated from molecular self-assembly in chemistry, but it is no longer limited to the molecular scale and has been widely used in MEMS, photonics, nano-science, near-field optics, and other fields. Since it is difficult for traditional assembling techniques to process complex structures and assemble high-volume components of different materials, the techniques are unable to meet the needs of efficiency and low cost. Therefore, the self-assembly technique has become a hot spot in the research of device and structure assembling technology.

At present, by using surface tension, scientists have achieved self-assembly for multiple batches. For example, automatically assembled LEDs have been realized on the appointed zone of substrates and have passed the electric test. Some Chinese scholars have also successfully achieved self-assembly by putting small squares of silicon on different substrates, such as nitride silicon and ordinary glass. Presently, they are researching self-assembling devices with non-symmetric and complex shape structures and are trying to manufacture some practicable structures using this technology. This self-assembling technology has a series of advantages, such as high accuracy and large assembly at a time, but the materials of devices and substrates can be inconsistent. At present, this technology has the most potential among MEMS integration technologies in terms of practicality.

The basic principles and steps of capillary-driven MEMS self-assembly are as follows:

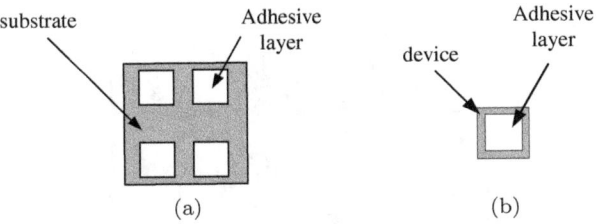

Figure 10.5. Diagram of device, substrate, and adhesive layer.

Table 10.1. The values of interface energy on different interfaces.

Interface	Surface energy (mJ·m^{-2})
Water — Adhesive	46
Water — Self-assembling membrane	52
Adhesive — Self-assembling hydrophobic membrane	1

1. Produce an assembled monolayer hydrophobic membrane as the adhesion layer with a special shape at the position of the substrate to install the device. Other parts of the substrate should remain hydrophilic, as shown in Figure 10.5(a).
2. Produce a self-assembly membrane with the corresponding shape of the fixing surface of the small device, as shown in Figure 10.5(b).
3. The surface of the substrate is coated with a special hydrophobic adhesive. When it is put into water, according to the minimum energy principle, the free energy of the interface always tends to its minimum. The adhesive will only attach to the hydrophobic membrane because of its hydrophobic characteristic and the relatively small interface energy between the adhesive and the self-assembled hydrophobic membrane, as shown in Table 10.1 and Figure 10.6.
4. A large number of micro-devices with adhesive are put into water and shaken continuously, as shown in Figure 10.7.
5. Once the hydrophobic parts of the device come into contact with the adhesive, they will automatically align with high precision and then complete the self-assembly through the solidifying process of the adhesive, as shown in Figure 10.8.

   The basic model is composed of three parts: the adhesive layer of the micro-device, the adhesive layer of the substrate, and the

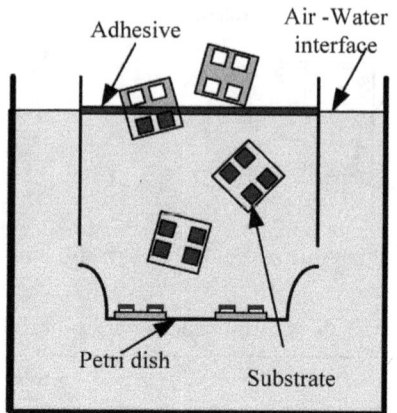

Figure 10.6.   Diagram of substrate coated with hydrophobic adhesive.

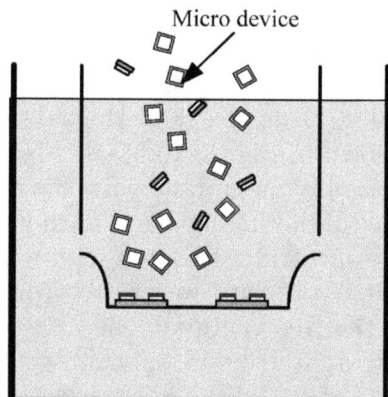

Figure 10.7.   Diagram of micro-device put into water.

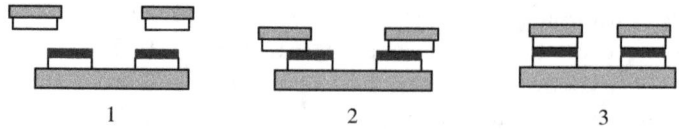

Figure 10.8.   Devices automatically assembled with substrate.

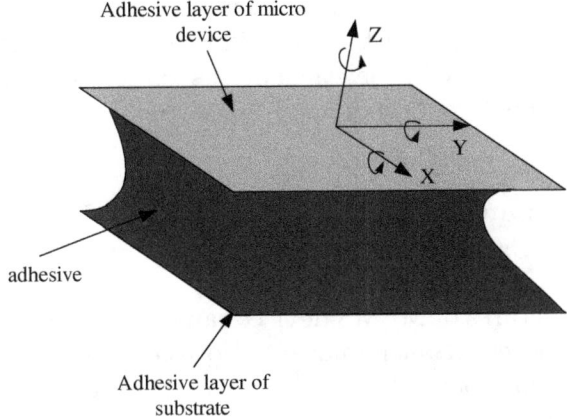

Adhesive layer of micro
device

adhesive

Adhesive layer of
substrate

Figure 10.9. Basic structure of model.

adhesive. The system has six degrees of freedom: three transla-
tional degrees of freedom and three rotational degrees of freedom
(Figure 10.9).

According to the basic law of thermodynamics, any closed system in nature is always consistent with the principle of minimum
energy. As a result, a liquid tends to assume the shape with the
minimum energy. In a mono-phase system, the surface area tends to
its minimum (most kinds of liquids take the shape of a ball without gravity). However, if there is an interface between the liquid
and another material, it will form a shape with a large surface. The
driving force of this process is the molecular interaction. Thus, the
difference between the surface tensions of different materials forms
the capillary force.

Self-assembly does not impose stringent criteria regarding the precision of the alignment between the dimensions and morphology of
the assembled device and the adhesive layer of the substrate. A design
layout of the substrate adhesive layer may be suitable for devices
with different sizes of the adhesive layer, so that one layout of the
substrate may be applied for the installation of more devices. On
the contrary, the devices with the same adhesive layer may be fixed
on different substrates. However, when fixing devices with different
shapes and sizes, the selection of the shape and size of the adhesive
layer should be considered. Of course, if the difference in the sizes
between adhesive layer and substrate adhesive layer of the device

is too large to meet the designed assembling precision, the determination of the limitation of the size difference of the adhesive layers during self-aligning relies on calculations under certain conditions, such as experiment conditions, and precision.

10.3 Dynamic Characteristics of Parallel-Plate Structure under Capillary Force

Most core structures of MEMS devices can be simplified into a model which is composed of an immovable bottom board and a movable parallel plate supported by a micro elastic beam. If the humidity in the environment where the MEMS is located is comparatively large, the liquid between the substrate and the movable plate will form a liquid bridge, as shown in Figure 10.10. The movable plate is generally driven by the sine of electrostatic force with known amplitude and frequency, so that it will form a dynamic liquid bridge between the movable plate and the substrate. It is necessary to analyze the dynamic characteristics of the parallel-plate structure under the electrostatic force and liquid bridge force because the dynamic characteristics of the parallel-plate structure have an influence on the performance of the whole MEMS.

The geometric structure of the liquid bridge between parallel plates is shown in Figure 10.11. In addition to the electrostatic force, the linear viscoelastic force, and the capillary force between the parallel plates will have an impact on the movable plate.

The total capillary force is the summation of the difference between the surface tension and the surface pressure of the meniscus.

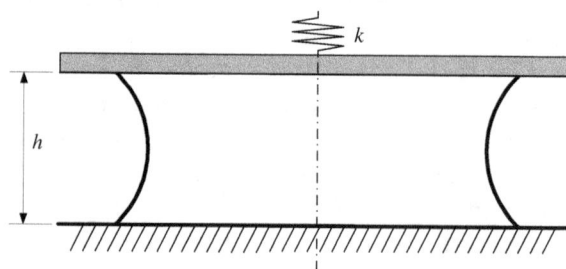

Figure 10.10. Liquid bridge between micro parallel plates.

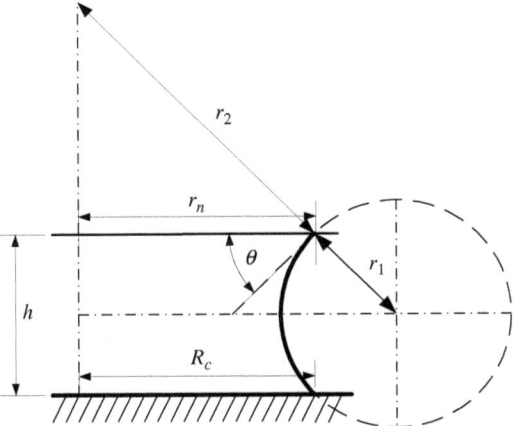

Figure 10.11. Geometric structure between parallel plates.

Its analytical expression is as follows:

$$F_{\textbf{cap}} = 2\pi r_n \gamma - \pi r_n^2 \gamma \left(\frac{1}{r_2} - \frac{1}{r_1} \right) \qquad (10.1)$$

where γ is the surface tension, r_n is the radius of the contour formed when the liquid bridge comes into contact with the oscillating plate, r_1 is the principal curvature radius of the liquid meniscus surface, and r_2 is the principal curvature radius of the cross-section of the liquid bridge and parallel plate.

The linear viscoelastic force of a liquid is a function of its dynamic viscosity of the liquid, which is expressed as follows:

$$F_{\textbf{vis}} = c\left(\eta^*\right) \dot{z} \qquad (10.2)$$

When all the electrostatic driving force, linear viscoelastic force of the liquid, gravity, elastic force of the spring, and capillary force have an impact on the movable plate, its equation can be expressed as follows:

$$m\frac{d^2 z}{dt^2} + c(\eta^*)\frac{dz}{dt} + kz + 2\pi r_n \gamma - \pi r_n^2 \gamma \left(\frac{1}{r_2} - \frac{1}{r_1} \right) = F_0 \sin \omega t \qquad (10.3)$$

where m is the mass of the movable plate, $c(\eta^*)$ is the damping coefficient of the liquid linear viscoelastic force, k is the stiffness

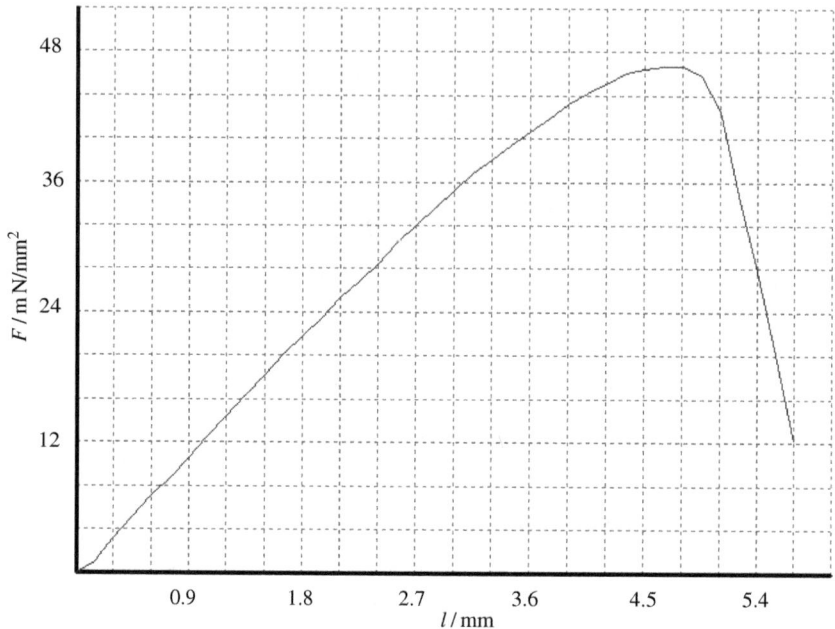

Figure 10.12. Measurement curve of liquid surface tension.

coefficient of the elastic supporting beam, z is the displacement of the movable plate, F_0 is the amplitude of the driving force, and ω is the frequency of the driving force.

In the experiment of using the Du Noüy ring method or a silicon film to measure the surface tension of water, the relation between pull and displacement is shown in Figure 10.12. The surface tension refers to the peak force, so that it has the meaning of strength.

When the displacement is small enough and the pull does not reach the peak (the ring or silicon film at the bottom), the value of pull will linearly decrease with the decreasing displacement. It is proved that water has an elastic characteristic under the condition of small deformation (not reaching the strength value of the surface tension).

In the water-stretching experiment of various areas of silicon films, the pull is proportional to the area of the silicon film, as shown in Figure 10.13. The experiment also indicates that for different areas, the curves of pull–displacement on unit area are coincident.

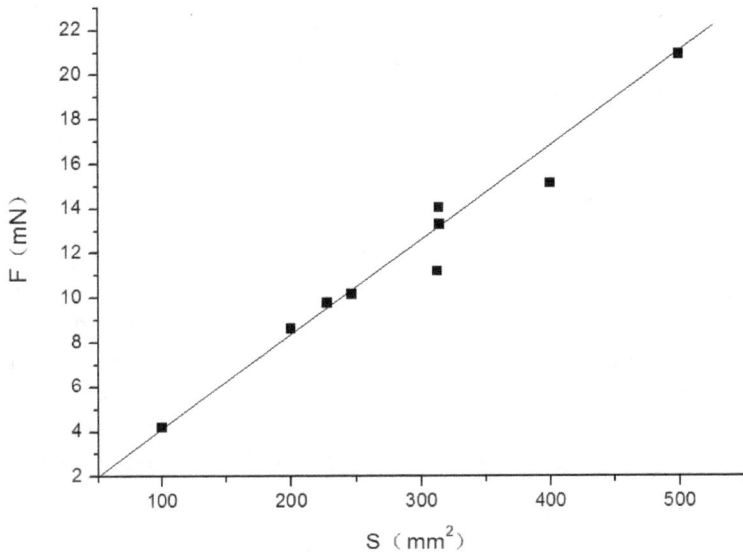

Figure 10.13. Relation curve between liquid surface tension and silicon film area.

It indicates that the capillary force is a linear function of the oscillation amplitude z, as given by

$$F_{\mathbf{cap}} = k_{\mathbf{cap}} z \qquad (10.4)$$

Since the curve in Figure 10.12 is a value of the pull force on unit area, $k_{\mathbf{cap}}$ is the product of the silicon area and the slope of an approximate straight line in the elastic section, as shown in Figure 10.12.

Substituting equation (10.4) into equation (10.3) leads to

$$m\frac{d^2 z}{dt^2} + c(\eta^*)\frac{dz}{dt} + kz - k_{\mathbf{cap}} z = F_0 \sin \omega t \qquad (10.5)$$

The solution to the equation above is

$$z = u \sin(\omega t + \varphi) \qquad (10.6)$$

The amplitude is

$$u = \frac{F_0}{k - k_{\mathbf{cap}}} \cdot \frac{1}{\sqrt{(1 - (\frac{\omega}{\omega_n})^2)^2 + (2 \cdot \xi \cdot \frac{\omega}{\omega_n})^2}} \qquad (10.7)$$

The phase is

$$\varphi = \arctan \left(\frac{2 \cdot \xi \cdot \frac{\omega}{\omega_n}}{1 - (\frac{\omega}{\omega_n})^2} \right) \tag{10.8}$$

where $\omega_n = \sqrt{\frac{k - k_{\mathbf{cap}}}{m}}$, $\xi = \frac{c(\eta^*)}{2\sqrt{(k - k_{\mathbf{cap}})m}}$.

If the displacement is comparatively large and the pull force of the water reaches the strength of surface tension, the elastic force will disappear and the force between the silicon piece and the water will be the capillary force, as shown in equation (10.1). Since the gap between the two plates is very small, r_2 is much larger than r_1. In this case, $\pi r_n^2 \gamma / r_2$ will tend to zero. $\pi r_n^2 \gamma / r_2 \to 0$. Therefore, $2\pi r_n \gamma$ can be ignored. According to Figure 10.11, it can be found that

$$r_1 \cos \theta = \frac{h}{2} \tag{10.9}$$

Then, equation (10.1) can be changed into

$$F_{\mathbf{cap}} = \frac{2A\gamma \cos \theta}{h} \tag{10.10}$$

The area is $A = \pi r_n^2$ for a circular plate.
So, equation (10.3) can be changed into

$$m\frac{d^2 z}{dt^2} + c(\eta^*)\frac{dz}{dt} + kz = F_0 \sin \omega t - \frac{k_2}{z + h_0} \tag{10.11}$$

where $k_2 = 2A\gamma \cos \theta$ and h_0 is the initial space between the two plates. $z = h - h_0$. In this case, the system is a nonlinear system of forced vibrations.

Before the MEMS device reaches the value of surface tension, the stiffness coefficient of the system will be enlarged, which is called the soft spring effect, due to the elastic action of the liquid. The amplitude–frequency and phase–frequency characteristics of the system will be changed. If $k_{\mathbf{cap}}$ is larger than k, the elastic restoring force is smaller than the pull force of the liquid. Thus, the system will adhere. Therefore, when designing the support beam in MEMS devices, the elastic restoring force should be larger than the pull force of the liquid.

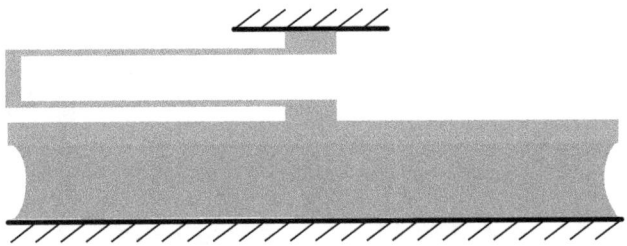

Figure 10.14. Liquid bridge between MEMS structures.

After the MEMS device reaches the value of surface tension, if the elastic restoring force is smaller than F_{cap}, the system will adhere. If the elastic restoring force is larger than F_{cap}, the system will exhibit nonlinear vibrations. The stability and sensitivity of the system will decrease.

To analyze the dynamic characteristics of the MEMS device under capillary force, the liquid bridge between the MEMS structures is shown in Figure 10.14.

The parameters of the system are as follows:

The mass of the microstructure is 0.243389×10^{-6} kg. The area of the microstructure is 0.55×10^{-6} m^2. The amplitude of the electrostatic driving force is 5.7657×10^{-6} N and its frequency is 4123 Hz. The initial space between the microstructures is 10 μm. In addition, the measurement value of the surface tension of the water is 73×10^{-3} N/m, and the contact angle is 79°.

Substituting the parameters mentioned above into equation (10.4), the elastic coefficient of the capillary force in the elastic section F_{cap} is obtained as 5.5 N/m. Therefore, if the stiffness of the microstructure is smaller than 5.5 N/m, the system will adhere. If the stiffness is larger than 5.5 N/m, the amplitude–frequency and phase–frequency characteristics of the system will be changed due to the soft spring effect. According to the analysis presented, the capillary force will severely affect the reliability, stability, and sensitivity of the micromachine structure. The capillary force must be eliminated during the process of manufacture, sealing, and storing of micromechanical devices.

Chapter 11

Capillary Wave

11.1 Phenomenon of Capillary Wave

A capillary wave is a liquid-free surface wave caused by capillary force. When the free surface is horizontal, the fluid is in a balance state. When some part of the fluid is disturbed, the particles of the liquid will leave the balance position and move along the direction which is perpendicular to the level surface. The particles will return to the balance position due to a restoring force, such as surface tension or gravity. Then, the particles move to the other side due to inertia. Thus, the vibration of the liquid particles is formed, and free surface waves are generated by the vibration transmission.

Generally, free surface waves are dominated by gravity and capillary force. Therefore, they include two components: capillary waves and gravity waves. However, on different scales, the effects of gravity and capillary force are different. Therefore, the types of surface waves are different due to the different wavelengths. If the wavelength is much larger than the wave height, the curvature of the free surface is very small, and the component of surface tension, which is perpendicular to the surface, is so small that it can be ignored. Only the gravity is considered. This kind of free surface wave is called a gravity wave. On the contrary, if the wavelength is extremely short, only surface tension is considered. Such a wave is called a capillary wave or ripple, whose wavelength typically ranges from one to two centimeters.

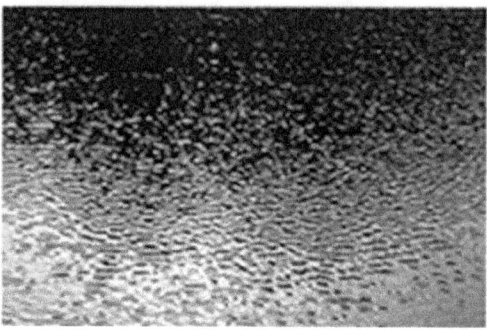

Figure 11.1.   Ripple.

Since the performances and mechanisms of gravity and capillary force are different, capillary waves and gravity waves have different characteristics. The shorter the wavelength of capillary waves, the higher the spread speed. However, the behavior of gravity waves is contrary to that of capillary waves. Capillary waves are common in nature. For example, ripples are a kind of capillary wave, as shown in Figure 11.1.

From the perspective of perturbation, the most common disturbance on the surface of water is wind. Wind generates waves, which is the most common phenomenon. At present, from the perspective of interference force, there are two main explanations for the generation of wind and waves. One is that due to the inconsistent distribution of wind, part of the surface of water is affected by pressure to generate waves. Once the waves have been produced, the wind force has a direct effect, and the waves continue to increase. The other explanation is based on the perspective of hydromechanics. When two kinds of fluids, which are in contact with each other, move relative to each other, the interface becomes unstable. It must form a wave surface to maintain the balance. When there are some little concave and convex areas on the surface of water, the airflow rates at the points of concave and convex surfaces are different. At the convex surface, the wind speed increases, and the wind pressure decreases, so that a rising trend is formed. At the concave surface, the wind speed decreases, and the pressure increases, so that a downward trend is formed. Therefore, the water surface became more uneven. The surface tension and gravity on the water surface produce a trend of restitution for the surface, which forms a storm.

If there is no wind, the sea remains calm. When the wind speed reaches a certain value but is still weak, capillary waves begin to appear on the water surface. The wavelength of capillary waves is so short and their wave height is so small that these waves can be seen as a wrinkle on the water surface. It only exists on a thin layer of the water surface. As the wind continues to blow, the wavelength and wave height of the capillary waves increase, and gravity waves are formed.

11.2 Dispersion Relation of the Free Surface Wave

The free surface wave has its frequency and a wavelength. Wavelength can be expressed by wave number. The relation between wave frequency and wave number is called the dispersion relation.

There is an interface tension at the interface of two kinds of fluids. When the upper-layer fluid is gas, it is called surface tension, whose direction is tangential to the interface of the fluid. In general, when the fluid is in a balance state, the free surface is horizontal. The free surface separates the upper and lower layers of the fluid. ρ' and ρ stand for the densities of the two kinds of fluid. For certain fluids, the densities are constant and $\rho > \rho'$. There is a hypothesis that the two kinds of fluid are not mucous, cannot be compressed, and move without eddies. The two layers of the fluid have a velocity potential, ϕ and ϕ'. The lower velocity and the higher velocity can be expressed by $\nabla\phi$ and $\nabla\phi'$, respectively.

There are three kinds of energy in a fluid system: gravity potential energy $V_{\mathbf{g}}$, caused by gravity; free surface energy $V_{\mathbf{st}}$, caused by surface tension; and kinetic energy T, caused by the flow. The potential energy $V_{\mathbf{g}}$ of gravity is easy to understand. $V_{\mathbf{g}}$ is determined by the density ρ, acceleration of gravity g, and height z from the surface, where $z = \eta(x, y, t)$. The potential energy of gravity can be obtained by volume integration in the directions of x, y, z, as

$$V_{\mathbf{g}} = \iint dx dy \int_0^{\eta} dz (\rho - \rho')gz = \frac{1}{2}(\rho - \rho')g \iint dx dy \eta^2 \quad (11.1)$$

where z is assumed to be zero at the even interface.

With an increase in the surface area, the energy will increase proportionally due to the effects of surface tension. Therefore, the surface free energy (potential energy of the surface tension) can be expressed as

$$V_{st} = \gamma \iint dxdy \left[\sqrt{1 + \left(\frac{\partial \eta}{\partial x}\right)^2 + \left(\frac{\partial \eta}{\partial y}\right)^2} - 1 \right]$$

$$\approx \frac{1}{2}\gamma \iint dxdy \left[\left(\frac{\partial \eta}{\partial x}\right)^2 + \left(\frac{\partial \eta}{\partial y}\right)^2 \right] \tag{11.2}$$

The term after the first equal sign represents the Gaspard Monge expression method. That after the second equal mark fits the case where the value of the derivative is very small and the surface is not so rough.

The kinetic energy of the fluid can be written as

$$T = \frac{1}{2} \iiint \rho v^2 dV \tag{11.3}$$

Using the relation $v^2 = \nabla\phi \cdot \nabla\phi = |\nabla\phi|^2$, the total kinetic energy of the upper and lower layers of the fluid can be written as

$$T = \frac{1}{2} \iint dxdy \left[\int_{-\infty}^{\eta} dz\rho |\nabla\phi|^2 + \int_{\eta}^{+\infty} dz\rho' |\nabla\phi'|^2 \right] \tag{11.4}$$

Since this kind of fluid is not compressed and has no eddies, $\phi(x, y, z, t)$ and $\phi'(x, y, z, t)$ should satisfy the Laplace equation:

$$\nabla^2\phi = 0 \quad \text{and} \quad \nabla^2\phi' = 0 \tag{11.5}$$

According to the first Green Formula, $\int_U (\psi\nabla^2\phi)dV = \oint_{\partial U} (\psi(\nabla\phi \cdot \vec{n}))dS - \int_U (\nabla\phi \cdot \nabla\psi)dV$, where $\nabla^2 = \Delta$ is the Laplacian, ∂U is the boundary of the zone U, and $\vec{n}$ is the normal vector of dS. If $\psi = \phi$, then

$$\int_U (\phi\nabla^2\phi)dV = \oint_{\partial U} (\phi(\nabla\phi \cdot \vec{n}))dS - \int_U (\nabla\phi \cdot \nabla\phi)dV \tag{11.6}$$

For the problem discussed here, since $\nabla^2\phi = 0$, the left-hand side of equation (11.6) is equal to 0. Then,

$$\oint_{\partial U} (\phi(\nabla\phi \cdot \vec{n}))dS = \int_U (\nabla\phi \cdot \nabla\phi)dV = \int_U |\nabla\phi|^2 \, dV \qquad (11.7)$$

Generally,

$$\nabla\phi = \frac{\partial\phi}{\partial x}\vec{i} + \frac{\partial\phi}{\partial y}\vec{j} + \frac{\partial\phi}{\partial z}\vec{k} \qquad (11.8)$$

Here, we only consider the locomotion along the direction of z. Therefore, $\nabla\phi = \partial\phi/\partial z \vec{k}$.

In addition, because the wave height is much smaller than the fluid height, the integration of equation (11.4) approximates to $\eta = 0$. The boundary of ∂U in equation (11.7) is shown in Figure 11.2.

For the lower layer of the fluid, when $z \to -\infty$, $\nabla\phi \to 0$ and $\nabla\phi \cdot \vec{n} = \partial\phi/\partial z \vec{k} \cdot \vec{n} = 0$. Therefore, in fact, only when $z = 0$, the integration along the boundary is not zero. Similarly, for the upper layer of the fluid, when $z \to +\infty$, $\nabla\phi' \to 0$ and $\nabla\phi' \cdot \vec{n} = 0$. Therefore, again, only when $z = 0$, the integration along the boundary is not zero. Then, equation (11.4) can be transformed into equation

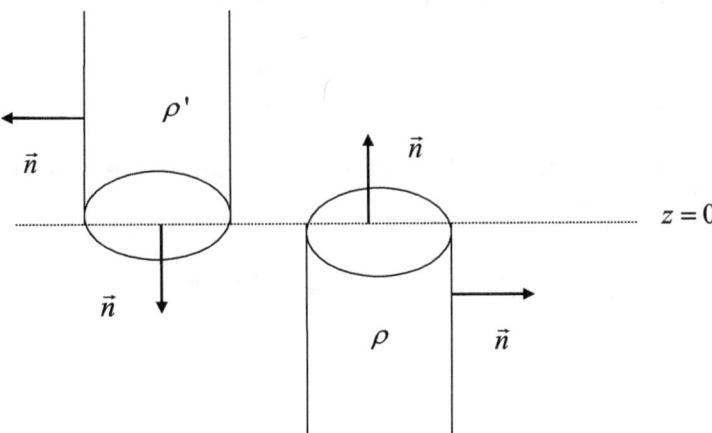

Figure 11.2. Boundaries of the upper and lower layers of the fluid.

(11.9) using the relation in (11.7):

$$T \approx \frac{1}{2} \iint dxdy \left(\rho\varphi \frac{\partial\varphi}{\partial z} - \rho'\varphi' \frac{\partial\varphi'}{\partial z} \right) \qquad (11.9)$$

In order to find the dispersion relation, consider the wave spreading along the direction of x. The equation of the wave is

$$\eta = a\cos(kx - \omega t) = a\cos\theta \qquad (11.10)$$

where a is the amplitude and k is the wave number. The phase is $\theta = kx - \omega t$. The velocity of the corresponding particle is $\partial\eta/\partial t$. Since the particle velocity expressed by the velocity potential is the same as the particle velocity on the boundary of $z = 0$,

$$\left.\frac{\partial\varphi}{\partial z}\right|_{z=0} = \frac{\partial\eta}{\partial t} \quad \text{and} \quad \left.\frac{\partial\varphi'}{\partial z}\right|_{z=0} = \frac{\partial\eta}{\partial t} \qquad (11.11)$$

When considering primarily the locomotion of the particle in the direction perpendicular to the liquid plane, the locomotion along the vertical direction and that along the horizontal direction are relatively independent. Therefore, assume that

$$\phi = \phi_z \cdot \phi_{xy} \qquad (11.12)$$

According to equation (11.5), $\nabla^2\phi = 0$. We can get

$$\nabla^2\phi = \frac{\partial^2\phi}{\partial x^2} + \frac{\partial^2\phi}{\partial y^2} + \frac{\partial^2\phi}{\partial z^2} = 0 \qquad (11.13)$$

Substituting equation (11.12) into partial differential equation (11.13) leads to

$$\nabla^2\phi = \frac{\partial^2\phi_z\phi_{xy}}{\partial x^2} + \frac{\partial^2\phi_z\phi_{xy}}{\partial y^2} + \frac{\partial^2\phi_z\phi_{xy}}{\partial z^2} = 0 \qquad (11.14)$$

That is,

$$\nabla^2\phi = \frac{\partial^2\phi_{xy}}{\partial x^2}\phi_z + \frac{\partial^2\phi_{xy}}{\partial y^2}\phi_z + \frac{\partial^2\phi_z}{\partial z^2}\phi_{xy} = 0 \qquad (11.15)$$

Dividing both sides of equation (11.15) by $\phi_z\phi_{xy}$ leads to

$$\left(\frac{\partial^2 \phi_{xy}}{\partial x^2} + \frac{\partial^2 \phi_{xy}}{\partial y^2}\right) \Big/ \phi_{xy} = -\frac{\partial^2 \phi_z}{\partial z^2} \Big/ \phi_z \tag{11.16}$$

When $z = 0$, $\partial\varphi/\partial z|_{z=0} = \partial\phi_z/\partial z|_{z=0} \cdot \phi_{xy} = \partial\eta/\partial t = a\omega\sin\theta$. Therefore, $\phi_{xy} = c_1 a\omega\sin\theta$, and we can get

$$\left(\frac{\partial^2 \phi_{xy}}{\partial x^2} + \frac{\partial^2 \phi_{xy}}{\partial y^2}\right) \Big/ \phi_{xy} = -k^2 \tag{11.17}$$

At the same time, using the relation in equation (11.16), it is known that

$$-\frac{\partial^2 \phi_z}{\partial z^2} \Big/ \phi_z = -k^2 \tag{11.18}$$

Therefore,

$$\frac{\partial^2 \phi_z}{\partial z^2} - k^2\phi_z = 0 \tag{11.19}$$

Finally, the general expression of ϕ_z is

$$\phi_z = Ae^{+|k|z} + Be^{-|k|z} \tag{11.20}$$

If $z \to -\infty$, ϕ is zero. Therefore, $B = 0$, and

$$\phi_z = Ae^{+|k|z} \tag{11.21}$$

When $z = 0$,

$$\frac{\partial\varphi}{\partial z} = \frac{\partial\phi_z}{\partial z} \cdot \phi_{xy} = A\,|k|\,e^{+|k|z} \cdot c_1 a\omega\sin\theta$$
$$= A\,|k|\,c_1 a\omega\sin\theta = a\omega\sin\theta \tag{11.22}$$

Therefore,

$$A = \frac{1}{c_1\,|k|} \tag{11.23}$$

Then,

$$\phi_z = +\frac{1}{|k|\,c_1}e^{+|k|z} \tag{11.24}$$

$$\phi(x, y, z, t) = \phi_z \cdot \phi_{xy} = +\frac{1}{|k|}e^{+|k|z}wa\sin\theta \qquad (11.25)$$

Similarly,

$$\phi'(x, y, z, t) = -\frac{1}{|k|}e^{-|k|z}wa\sin\theta \qquad (11.26)$$

Substituting equations (11.25) and (11.26) into equation (11.9), integrating the equations (11.1), (11.2), and (11.9) along the x and y directions, whose integration ranges are one wavelength, $\lambda = 2\pi/k$, in the x direction and one unit length in the y direction, leads to

$$V_{\mathbf{g}} = \frac{1}{4}(\rho - \rho')ga^2\lambda \qquad (11.27)$$

$$V_{\mathbf{st}} = \frac{1}{4}\gamma k^2 a^2\lambda \qquad (11.28)$$

$$T = \frac{1}{4}(\rho + \rho')\frac{\omega^2}{|k|}a^2\lambda \qquad (11.29)$$

According to the Lagrange function, L is defined as $L = T - V$. V is the algebraic sum of the gravity potential energy $V_{\mathbf{g}}$ and the potential energy of the surface tension $V_{\mathbf{st}}$. Then,

$$L = \frac{1}{4}\left[(\rho + \rho')\frac{\omega^2}{|k|} - (\rho - \rho')g - \lambda k^2\right]a^2\lambda \qquad (11.30)$$

Substituting this into the Euler–Lagrange equation,

$$\frac{d}{dt}\left(\frac{\partial L}{\partial \dot{x}_i}\right) - \frac{\partial L}{\partial x_i} = 0 \qquad (11.31)$$

The variable of the generalized coordinates is only a. We can get

$$\omega^2 = |k|\left(\frac{\rho - \rho'}{\rho + \rho'}g + \frac{\gamma}{\rho + \rho'}k^2\right) \qquad (11.32)$$

where g is the acceleration, ρ and ρ' are the mass densities of the two kinds of fluids, with $\rho > \rho'$, and $\rho - \rho'/\rho + \rho'$ in the first term is called the Atwood number. This expresses the relation between frequency and number of the wave (or wavelength), which is called the dispersion relation.

11.3 Analysis of the Characteristics of Free Surface Waves and the Concept of Capillary Waves

Wave velocity is the transmission rate of a wave. According to the different characteristics of waves, the definitions of wave velocity are different, such as phase velocity, group velocity, signal velocity, and front velocity. Generally, wave velocity refers to phase velocity, which is discussed in this section. The phase velocity of a wave is the rate at which the phase of the wave transfers in space. This is the velocity at which the phase of any frequency component of the wave travels. Any given phase of the wave (for example, the peak) will travel at the phase velocity, as shown in Figure 11.3.

As an example of the phase velocity of a wave peak, the corresponding phase is $\theta = 0$. Assuming that the wave peak reaches x_1 at time t_1 and reaches x_2 at time t_2, since $kx_1 - \omega t_1 = 0$ and $kx_2 - \omega t_2 = 0$, the transformation distance per unit time is the phase velocity of the wave peak, which can be calculated as follows:

$$C = \frac{x_2 - x_1}{t_2 - t_1} = \frac{\omega}{k} \tag{11.33}$$

Therefore, C can be expressed using angle frequency ω and wave number k.

In addition, since $2\pi/k$ reflects the wavelength, the length of wave in one period is $\lambda = 2\pi/k$, so $k = 2\pi/\lambda$. Substituting it into the

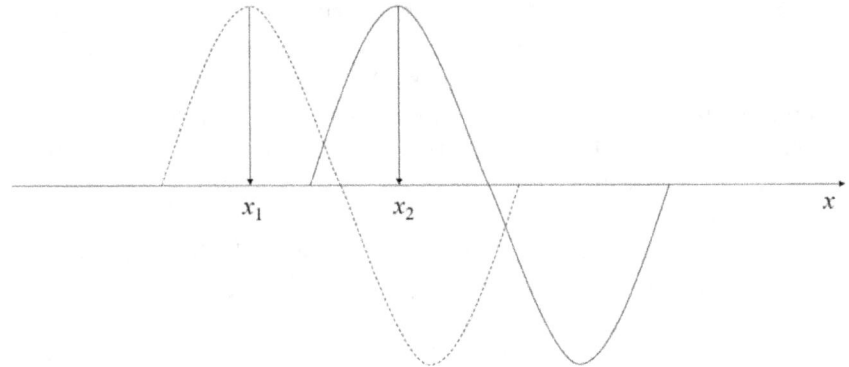

Figure 11.3. Diagram of the transformation of wave peak.

dispersion expression (11.32), it can be deduced that

$$C^2 = \frac{g\lambda}{2\pi}\frac{\rho - \rho'}{\rho + \rho'} + \frac{2\pi\gamma}{\lambda(\rho + \rho')} \qquad (11.34)$$

where C is the wave velocity, g is the acceleration due to gravity, λ is the wavelength, ρ' and ρ are the densities of the upper and lower layers of the fluid, and γ is the surface tension.

The first term on the right-hand side of equation (11.34) includes g, which reflects the effect of gravity. The second term includes γ, which reflects the effect of the surface tension. For waves with very short wavelengths, the smaller the wavelength, the greater the influence of the surface tension on wave velocity. For waves with very long wavelengths, the larger the wavelength, the greater the effect of gravity on the wave velocity. When the wavelength is equal to

$$\lambda_c = 2\pi\sqrt{\frac{\gamma}{(\rho - \rho')g}} \qquad (11.35)$$

the wave velocity has a minimum value of $C_{\min}$, and

$$C_{\min}^2 = 2\frac{\rho - \rho'}{\rho + \rho'}\sqrt{\frac{g\gamma}{\rho - \rho'}} \qquad (11.36)$$

For short waves, since the wavelength is comparatively short, $\lambda < \lambda_c$, the second term on the right-hand side of equation (11.34) is dominant. The role of surface tension is remarkable. Kelvin suggested that this kind of wave should be called a "ripple", which is also known as a capillary wave. For capillary waves, the shorter the wavelength, the greater the velocity. The wave velocity decreases with an increase in the wavelength. When $\lambda = \lambda_c$, surface tension and gravity work on the fluctuations at the same time. This wave is called a capillary-gravity wave. Only when $\lambda > \lambda_c$, the effect of surface tension reduces. In this case, fluctuations are mainly caused by gravity, and the wave is called a gravity wave.

For a two-layer fluid which is composed of air and water, if the surface tension is $\gamma = 0.073\,\mathrm{N/m}$, the density of water is $\rho = 1000\,\mathrm{kg/m^3}$, the density of air is $\rho' = 1.299\,\mathrm{kg/m^3}$, and $g = 9.8\,\mathrm{m/s^2}$, then the wavelength is $\lambda_{\min} = 1.78\,\mathrm{cm}$ at the minimum wave velocity. The minimum wave velocity is $C_{\min} = 23.2\,\mathrm{cm/s}$.

Bibliography

[Anon.] (1911). Capillary action. In *Encyclopedia Britannica*.

Aaronson, S. (2005). NP-complete problems and physical reality. *SIGACT News*, 36(1), 30–52.

Adam, N. K. (1941). *The Physics and Chemistry of Surfaces* (3rd ed.). Oxford University Press.

Adamson, A. W., and Gast, A. P. (1997). *Physical Chemistry of Surfaces* (6th ed.). Wiley.

Baierlein, R. (2003). *Thermal Physics*. Cambridge University Press.

Ball, W. W. R. (2003). Pierre Simon Laplace (1749–1827). In *A Short Account of the History of Mathematics* (4th ed.). Dover.

Bennett, C. O., and Myers, J. E. (1962). *Momentum, Heat, and Mass Transfer*. McGraw-Hill.

Bhushan, B. (1999). *Principles and Applications of Tribology*. Wiley.

Bormashenko, E. (2008). Why does the Cassie-Baxter equation apply? *Colloids Surf. A*, 324, 47–50. Lin, F. (2008). Petal effect: Two major examples of the Cassie-Baxter model are the petal effect and lotus effect — A superhydrophobic state with high adhesive force. *Langmuir*, 24(8), 4114–4119.

Bowden, F. P., and Tabor, D. (1954). *Friction and Lubrication of Solids, Part I*. Oxford University Press.

Bowden, F. P., and Tabor, D. (1964). *Friction and Lubrication of Solids, Part II*. Oxford University Press.

Boys, C. V. (1911). *Soap Bubbles: Their Colours and the Forces That Mould Them* (2nd ed.).

Bush, J. W. M. (2004a). MIT lecture notes on surface tension, lecture 1. Massachusetts Institute of Technology.

Bush, J. W. M. (2004b). MIT lecture notes on surface tension, lecture 3. Massachusetts Institute of Technology.

Bush, J. W. M. (2004c). MIT lecture notes on surface tension, lecture 5. Massachusetts Institute of Technology.

Bush, J. W. M. (2004d). Surface tension module. MIT OpenCourseWare.

Cabezas, G., Montanero, J. M., Acero, J., Jaramillo, M. A., and Fernández, J. A. (2002). Detection of liquid bridge contours and its applications. *Meas. Sci. Technol.*, 13, 725–732.

Calvert, J. B. (2007). Surface tension (physics lecture notes). University of Denver.

Chau, A., Regnier, S., Delchambre, A., and Lambert, P. (2007). Influence of geometrical parameters on capillary forces. In *Proceedings of the 2007 IEEE International Symposium on Assembly and Manufacturing* (pp. 215–220).

Chow, T. S. (1998). Wetting of rough surfaces. *J. Phys. Condens. Matter*, 10(27), L445–L451.

Christenson, H. K. (1988). Adhesion between surfaces in unsaturated vapors: A reexamination of the influence of meniscus curvature and surface forces. *J. Colloid Interface Sci.*, 121(1), 170–178.

Christenson, H. K., and Yaminsky, V. V. (1993). Adhesion and salvation forces between surfaces in liquids studied by vapor-phase experiments. *Langmuir*, 9, 2448–2454.

Comyn, J. (1997). *Adhesion Science*. Royal Society of Chemistry.

Dataphysics (2007). Sessile drop method. Technical Documentation.

de Gennes, P. G. (1985). Wetting: Statics and dynamics. *Rev. Mod. Phys.*, 57(3), 827–863.

de Gennes, P. G. (1994). *Soft Interfaces*. Cambridge University Press.

de Gennes, P. G., Brochard-Wyart, F., and Quéré, D. (2002). *Capillary and Wetting Phenomena — Drops, Bubbles, Pearls, Waves*. Springer.

Dean, J. A. (Ed.). (1973). *Lange's Handbook of Chemistry* (10th ed.). McGraw-Hill.

Dingemans, M. W. (1997). *Water Wave Propagation Over Uneven Bottoms*. World Scientific.

Ertl, G., Knözinger, H., and Weitkamp, J. (1997). *Handbook of Heterogeneous Catalysis* (Vol. 2). Wiley-VCH.

Eustathopoulos, N., Nicholas, M. G., and Drevet, B. (1999). *Wettability at High Temperatures*. Pergamon.

Fan, H., and Gao, Y. X. (2001). Elastic solution for liquid-bridging-induced microscale contact. *J. Appl. Phys.*, 90(12), 5904–5910.

Fan, H., and Wang, G. F. (2003). Stability analysis for liquid-bridging induced contact. *J. Appl. Phys.*, 93(5), 2554–2558.

Ferraro, P. (1998). What breaks the shadow of the tube? *Phys. Teach.*, 36, 542–543.

Ferraro, P. (2008). Wettability patterning of lithium niobate substrate by modulating pyroelectric effect to form microarray of sessile droplets. *Appl. Phys. Lett.*, 92, 213107.

Finn, R. (1999). *Capillary Surface Interfaces.* American Mathematical Society.

Fisher, L. R., and Israelachvili, J. N. (1981). Direct measurement of the effect of meniscus forces on adhesion: A study of the applicability of macroscopic thermodynamics to microscopic liquid interfaces. *Colloids Surf.*, 3, 303–319.

Gao, C. (1997). Theory of menisci and its applications. *Appl. Phys. Lett.*, 71, 1801–1803.

Gao, S., Jin, L., Du, J., and Liu, H. (2011). The liquid-bridge with large gap in micro structural systems. *J. Mod. Phys.*, 2, 404–415.

Gao, S., and Liu, H. (2008). *Mechanics of Micro-Electrical-Mechanism Systems.* National Defense Industry Press.

Goljan, E. (2007). *Pathology* (2nd ed.). Mosby Elsevier.

Good, R. J. (1992). Contact angle, wetting, and adhesion: A critical review. *J. Adhes. Sci. Technol.*, 6(12), 1269–1302.

Gregg, S. J., and Sing, K. S. W. (1982). *Adsorption, Surface Area and Porosity* (2nd ed.). Academic Press.

Harrison, C. (Ed.). (1960). *Handbook of Chemistry and Physics* (42nd ed.). CRC Press.

He, M., Blum, A. S., Aston, D. E., Buenviaje, C., and Overney, R. M. (2001). Critical phenomena of water bridges in nanoasperity contacts. *J. Chem. Phys.*, 114(3), 1355–1360.

International Union of Pure and Applied Chemistry Commission on Atmospheric Chemistry. (1990). Glossary of atmospheric chemistry terms. *Pure Appl. Chem.*, 62, 2167–2219.

International Union of Pure and Applied Chemistry Commission on Physicochemical Symbols Terminology and Units. (1993). *Quantities, Units and Symbols in Physical Chemistry* (2nd ed.). Blackwell Scientific Publications.

International Union of Pure and Applied Chemistry Commission on Quantities and Units in Clinical Chemistry, & International Federation of Clinical Chemistry Committee on Quantities and Units. (1996). Glossary of terms in quantities and units in clinical chemistry (IUPAC-IFCC Recommendations 1996). *Pure Appl. Chem.*, 68, 957–1000. https://doi.org/10.1351/pac199668040957.

Israelachvili, J. (2004). *Intermolecular and Surface Forces* (3rd ed.). Academic Press.

Jang, J., Schatz, G. C., and Ratner, M. A. (2004). Capillary force in atomic force microscopy. *J. Chem. Phys.*, 120(3), 1157–1160.

Johnson, R. E. (1993). Wettability. In J. C. Berg (Ed.), *Surface Chemistry and Applications*. Marcel Dekker.

Jurin, J. (1719). An account of some experiments shown before the Royal Society. *Philos. Trans. R. Soc. Lond.*, 30, 739–747.

Katchalsky, A., and Curran, P. F. (1965). *Nonequilibrium Thermodynamics in Biophysics*. Harvard University Press.

Kinloch, A. J. (1987). *Adhesion and Adhesives: Science and Technology*. Chapman and Hall.

Kirkness, J. P., Christenson, H. K., Wheatley, J. R., and Amis, T. C. (2005). Application of the 'pull-off' force method for measurement of surface tension of upper airway mucosal lining liquid. *Physiol. Meas.*, 26, 677–688.

Lamb, H. (1928). *Statics, Including Hydrostatics and the Elements of the Theory of Elasticity* (3rd ed.). Cambridge University Press.

Lamb, H. (1994). *Hydrodynamics* (6th ed.). Cambridge University Press.

Lambert, P., Seigneur, F., Koelemeijer, S., and Jacot, J. (2006). A case study of surface tension gripping: The watch bearing. *J. Micromech. Microeng.*, 16(7), 1267–1276.

Langmuir-Blodgett Instruments (2007). Surface and interfacial tension. Technical Documentation.

Laplace, P. S. (1806). *Mécanique Céleste, Supplement to the Tenth Edition*.

Lauda (2007). Surfactants at interfaces. Technical Documentation.

Lee, K. S. (2008). Kinetics of wetting and spreading by aqueous surfactant solutions. *Adv. Colloid Interface Sci.*, 144, 54–65.

Leenaars, A. F. M., Huethorst, J. A. M., and van Oekel, J. J. (1990). Marangoni drying: A new extremely clean drying process. *Langmuir*, 6, 1701–1703.

Marangoni, C. (1865). On the expansion of a drop of liquid floating in the surface of another liquid. Published Dissertation.

Marmur, A. (1992). Modern approach to wettability: Theory and applications. In M. E. Schrader & G. Loeb (Eds.), *Wettability*. Plenum Press.

Marmur, A. (2003). Wetting of hydrophobic rough surfaces: To be heterogeneous or not to be. *Langmuir*, 19, 8343–8348.

McFarlane, J. S., and Tabor, D. (1950). *Proc. R. Soc. Lond. A* (pp. 202–224).

McGraw-Hill. (2003). Jurin rule. In *McGraw-Hill Dictionary of Scientific and Technical Terms*.

Mendoza, E. (1988). *Reflections on the Motive Power of Fire — and Other Papers on the Second Law of Thermodynamics by E. Clapeyron and R. Carnot*. Dover Publications.

Moore, W. J. (1962). *Physical Chemistry* (3rd ed.). Prentice Hall.

Müller, I. (2007). *A History of Thermodynamics — The Doctrine of Energy and Entropy*. Springer.

Okumura, K. (2008). Wetting transitions on textured hydrophilic surfaces. *Eur. Phys. J. E*, 25, 415–424.

Orr, F. M., Scriven, L. E., and Rivas, A. P. (1975). Pendular rings between solids: Meniscus properties and capillary force. *J. Fluid Mech.*, 67, 723–742.

Oxford University Press. (1989). Formula. In *The Oxford English Dictionary* (2nd ed.).

Pakarinen, O. H., Foster, A. S., Paajanen, M., Kalinainen, T., Katainen, J., Makkonen, I., Lahtinen, J., and Nieminen, R. M. (2005). Towards an accurate description of the capillary force in nanoparticle-surface interactions. *Model. Simul. Mater. Sci. Eng.*, 13(7), 1175–1186.

Perrot, P. (1998). *A to Z of Thermodynamics*. Oxford University Press.

Petrovic, S., Robinson, T., and Judd, R. L. (2004). Marangoni heat transfer in subcooled nucleate pool boiling. *Int. J. Heat Mass Transfer*, 47(23), 5115–5128.

Pfitzner, J. (1976). Poiseuille and his law. *Anaesthesia*, 31(2), 273–275.

Phillips, O. M. (1977). *The Dynamics of the Upper Ocean* (2nd ed.). Cambridge University Press.

Physical Properties Sources Index: Eötvös Constant. (2008).

Phywe (2007). Surface tension by the ring method (Du Nouy method). Technical Documentation.

Quere, D. (2008). Wetting of textured surfaces. *Colloids Surf.*, 206, 41–46.

Reiss, H. (1965). *Methods of Thermodynamics*. Dover Publications.

Rowlinson, J. S., and Widom, B. (1982). *Molecular Theory of Capillarity*. Clarendon Press.

Safran, S. (1994). *Statistical Thermodynamics of Surfaces, Interfaces, and Membranes*. Addison-Wesley.

Salzman, W. R. (2001). Open systems. In Chemical thermodynamics. University of Arizona.

Schrader, M. E., and Loeb, G. I. (1992). *Modern Approaches to Wettability: Theory and Applications*. Plenum Press.

Scriven, L. E., and Sternling, C. V. (1960). The Marangoni effects. *Nature*, 187, 186–188.

Sears, F. W., and Zemanski, M. W. (1955). *University Physics* (2nd ed.). Addison Wesley.

Sharfrin, E. (1960). Constitutive relations in the wetting of low energy surfaces and the theory of the retraction method of preparing monolayers. *J. Phys. Chem.*, 64(5), 519–524.

Sherwood, L. (2007). Human physiology from cells to systems. In P. Adams (Ed.), *Human Physiology* (6th ed.). Thomson Brooks/Cole.

Shinto, H., Uranishi, K., Miyahara, M., and Higashitani, K. (2002). Wetting-induced interaction between rigid nanoparticle and plate: A Monte Carlo study. *J. Chem. Phys.*, 116(21), 9500–9509.

Stifter, T., Marti, O., and Bhushan, B. (2000). Theoretical investigation of the distance dependence of capillary and van der Waals forces in scanning force microscopy. *Phys. Rev. B*, 62(20), 13667–13673.

Sutera, S. P., and Skalak, R. (1993). The history of Poiseuille's law. *Annu. Rev. Fluid Mech.*, 25, 1–19.

Tadmor, R. (2004). Line energy and the relation between advancing, receding and Young contact angles. *Langmuir*, 20, 7659–7664.

Thomson, J. (1855). On certain curious motions observable on the surfaces of wine and other alcoholic liquors. *Philos. Mag.*, 10, 330–333.

Thomson, W. T. (1871). On capillary attraction. *Philos. Mag.*, 42, 448–452.

Tufillaro, N. B., Ramshankar, R., and Gollub, J. P. (1989). Order-disorder transition in capillary ripples. *Phys. Rev. Lett.*, 62(4), 422–425.

Van Krevelen, D. W. (1976). *Properties of Polymers* (2nd ed.). Elsevier.

Washburn, E. W. (1921). The dynamics of capillary flow. *Phys. Rev.*, 17(3), 273–283.

White, H. E. (1948). *Modern College Physics*. van Nostrand.

Whyman, G. (2008). The rigorous derivation of Young, Cassie–Baxter and Wenzel equations and the analysis of the contact angle hysteresis phenomenon. *Chem. Phys. Lett.*, 450, 355–359.

Wu, D., Fang, N., Sun, C., and Zhang, X. (2006). Stiction problems in releasing of 3d microstructures and its solution. *Sens. Actuat. A*, 128, 109–115.

Young, T. (1805). An essay on the cohesion of fluids. *Philos. Trans. R. Soc. Lond.*, 95, 65–87.

Yuan, J. Y., Shao, Z., and Gao, C. (1991). Alternative method of imaging surface topologies of nonconducting bulk specimens by scanning tunneling microscopy. *Phys. Rev. Lett.*, 67(7), 2901–2904.

Zhang, B. and Nakajima, A. (1999). Nanometer deformation caused by the Laplace pressure and the possibility of its effect on surface tension measurements. *J. Colloid Interface Sci.*, 211, 114–121.

Index